Essential Skills

and Strategies

in the Helping Process

Essential Skills and Strategies in the Helping Process

ROBERT E. DOYLE
ST. JOHN'S UNIVERSITY

BROOKS/COLE PUBLISHING COMPANY
PACIFIC GROVE, CALIFORNIA

 A CLAIREMONT BOOK

Brooks/Cole Publishing Company
A Division of Wadsworth, Inc.

Printed in the United States of America

10   9   8   7   6   5   4   3   2   1

Library of Congress Cataloging in Publication Data

Doyle, Robert E. [date]
    Essential skills and strategies in the helping process / by Robert
E. Doyle.
      p.   cm.
    Includes bibliographical references and indexes.
    ISBN 0-534-17328-4 (hardcover)
    1. Counseling   I. Title.
BF637.C6D63   1992   91-24852
158'.3—dc20   CIP

Sponsoring Editor *Claire Verduin*
Marketing Representative *Gerry Levine*
Editorial Associate *Gay C. Bond*
Production Editor *Kay Mikel*
Manuscript Editor *Nancy Hopkins*
Permissions Editor *Carline Haga*
Interior Design *Vernon T. Boes*
Cover Design *Cloyce Wall*
Art Coordinator *Lisa Torri*
Typesetting *Bookends Typesetting*
Cover Printing *Phoenix Color Corporation*
Printing and Binding *Arcata Graphics (Fairfield)*

To my wife, Madeleine,
and my daughters, Jeanne and Kristina,
for their infinite patience,
encouragement, and love.

Preface

The purpose of this book is to provide students who are studying the counseling process with a comprehensive understanding of the skills and strategies that are of critical importance in this process. The book is intended for students who are enrolled in an introductory counseling course at the graduate or the undergraduate level. It assumes that the reader has had little or no background in counseling theory or methodology. Although the book specifically addresses students who intend to become counselors in schools, colleges, and various community agencies, it may be used in other departments, such as human services, nursing, psychology, and social work, which offer courses in counseling.

A certain point of view is clearly implicit in any counseling textbook. This book evolved over several years from my interest in presenting my students with an understanding of the counseling process from a practical and skilled-based point of view. I have attempted to do several things: First, I wanted to present students with an overview of the components of the counseling process. Second, I wanted to provide students with an introduction to some of the more widely used counseling intervention strategies. Third, I wanted to help students master the communication skills that are essential to any counseling process. Finally, I wanted to furnish students with a basic resource book that they could use as a foundation for developing their knowledge and skills of the helping process. I have organized the book into three major sections.

In Part 1 of the book I try to help the student understand some of the essential concepts concerning the counselor, the client, and the counseling process. This part contains three chapters. In Chapter 1 I discuss important issues related to the education and training of the counselor. These include the counselor's personal qualities, the educational requirements for professional counselors, the major professional associations concerned with the counseling process, the ethical standards that counselors should be familiar with, and the legal concerns that affect the counseling process. In Chapter 2 I provide students with a brief review of some of the major reasons why individuals function with different degrees of effectiveness. The information covered in this chapter is not usually found in introductory counseling textbooks. I decided to incorporate this material because I found that many of my students needed both a review of these basic principles and a foundation for examining client concerns. In Chapter 3 I present students with a model for conceptualizing and understanding the counseling process as it unfolds or develops through time. In this chapter I also point out the difficulties encountered in working with both reluctant and resistant clients and suggest some ways that this process can be facilitated.

In Part 2 my goal is to enable students to gain a practical overview of some of the fundamental intervention strategies that are used by counselors to help clients function more effectively. This part of the textbook contains four chapters. In the introduction to this section of the text, I outline some of the forces that are attempting to integrate various counseling approaches and then present a skills-oriented model for conceptualizing various counseling approaches as intervention strategies rather than theories. In Chapter 4 I discuss four major cognitively oriented counseling strategies from this skills-oriented point of view. These four strategies are helping clients acquire and retain important factual information; assisting clients with the decision-making process; helping clients restructure their logical thinking; and assisting clients with analogous, inductive, and creative thinking. This chapter includes some interventions that are used extensively by counselors in a variety of settings but rarely found in counseling theory texts. In Chapter 5 I describe a number of affectively oriented approaches and then make some suggestions about how these approaches can be used to help clients ventilate, obtain catharsis, gain insight, and improve their self-concepts. In Chapter 6 I present the important ideas related to the behaviorally oriented or performance-based counseling strategies that have proved to be very useful in helping clients.

In Part 3 my objective is to provide students with a process that will enable them to develop the critical communication skills required for effective helping. The introduction to this section of the book outlines the basic communication roles employed by counselors and describes a four-point Likert-type scale for the qualitative evaluation of the counselor's competency and timing in using a particular role. This section of the text contains four chapters. In each chapter I describe certain role communication skills; outline the imporant subsidiary verbal responses that are associated with each of these roles; furnish illustrations of these communication skills; and provide numerous examples and practice exercises for developing these skills. In Chapter 7 I present the primary communication skills: attending, clarifying, and supporting. In Chapter 8 I discuss the intermediate communication skills: providing information, probing, and the use of silence. In Chapter 9 I describe the advanced communication skills: motivating or prescribing, and analyzing or evaluating. In Chapter 10 I provide students with the opportunity to practice communicating various intervention strategies.

This book was written with students in mind. The first six chapters may be read independently by students. The information provided in these chapters may be further developed by classroom discussions or supplementary lectures that can go into any of these topics in greater detail. The communication skills discussed in the last four chapters are designed to be used in a skill-building course. The exercises in these chapters need to be studied under the direction of a skilled counselor. I have tried to present the subject matter in a relatively informal writing style. The chapters were pretested with my own students, who were asked to provide feedback and pose questions about any concepts that were not completely understood.

ACKNOWLEDGMENTS

Developing this book has been a challenge, and I would like to express my appreciation to all those who have contributed to it both directly and indirectly.

First, I must thank the students who have taken counseling courses with me. They have been a constant source of inspiration, and I am indebted to them for the feedback and assistance they gave me in refining the materials and ideas presented.

Second, I want to thank those with whom I have been associated at St. John's University: Andrew Gentile, Shirley Griggs, Patricia Hudson, John Kukor, Donald Sampson, James O'Toole, William Swan, William Pilkington, and Nancy Kessler. They provided encouragement and many helpful suggestions as this project developed.

Third, special thanks are extended to those who reviewed preliminary drafts of parts of this manuscript and made insightful and constructive comments: Angelo V. Boy, University of New Hampshire; Kim C. Francis, Brooklyn College; Eugene Goldin, New York Institute of Technology; Anita Leach McMahon, University of South Florida; Rie Rogers Mitchell, California State University at Northridge; Margaret Olsen, University of Wisconsin at Oshkosh; Virginia Allen, Idaho State University; Stanley Baker, Penn State University; Fred Borgen, Iowa State University; Sam Gladding, Wake-Forest University; Mary F. Maples, University of Nevada; Bernard Nisenholz, California State University at Northridge; Thomas Skovholt, University of Minnesota; Brent Snow, Oklahoma State University; and Jane S. O'Hern, Boston University.

Finally, I want to express my gratitude to my editor, Claire Verduin, and to Gay C. Bond, Kay Mikel, Vernon Boes, and Lisa Torri at Brooks/Cole. They have been a pleasure to work with, and their support, efforts, and dedication have made this project a reality.

ROBERT E. DOYLE

Contents

CHAPTER 5

Affectively Focused Counseling Strategies   *94*

CHAPTER 6

Performance Focused Counseling Strategies   *117*

PART 3
The Counselor's Role Communication Skills 147

CHAPTER 7
Primary Role Communication Skills 157

The Supporting or Reassuring Role   179

Summary   191
References and Suggested Readings   192

Intermediate Role Communication Skills   193

Introduction   193
The Informing or Describing Role   193

The Inquiring or Probing Role   209

The Meaning and Use of Silence   223
Summary   226
References and Suggested Readings   227

CHAPTER 9

Advanced Role Communication Skills   228

Introduction   228
The Motivating or Prescribing Role   228

CHAPTER 10

Problem-Solving Skills   261

PART 1

The Counselor, the Client, and the Counseling Process

Counseling is a dynamic process during which a trained individual assists another person who is functioning at a low or ineffective level to become a more effectively functioning person. This part of the text discusses three unique factors involved in this definition: the counselor, the client, and the counseling process. The personal characteristics, professional education, and training of the counselor, and the ethical standards and legal considerations that guide the counselor's behavior in practice are of crucial importance in this process. These issues are discussed in Chapter 1. Individuals who come to a counselor for help may be functioning inappropriately in one or more major areas of life. Chapter 2 reviews how effective individuals function differently from ineffective persons in important components of life and discusses the areas where clients frequently need assistance. Chapter 3 delineates the characteristics of the counseling process as it progresses through different developmental phases or stages of counseling, and it outlines the distinctive activities that occur in each of these stages. This chapter also discusses some ways of dealing effectively with reluctant and resistant clients.

1

The Counselor and the Counseling Process

INTRODUCTION

The counselor's personal characteristics, professional education and training, and knowledge of important ethical and legal issues are the major topics presented in this chapter. After studying this chapter you should be able to:

- indicate the important characteristics of professional counselors;
- describe the professional education and training of counselors;
- name the major counseling associations;
- identify the major ethical standards or codes of conduct that guide practitioners in the counseling field;
- discuss four important ethical issues in counseling: the client's welfare, the competency of the counselor, the client's right to privacy, and value issues inherent in the counseling relationship; and
- specify some of the major legal issues that affect the counselor.

CHARACTERISTICS OF THE PROFESSIONAL COUNSELOR

The counselor is a trained professional person who should manifest the following personal and professional characteristics: (1) the belief that clients are unique individuals of significant value, (2) the belief that clients are capable of change, (3) the knowledge of how effective individuals function, (4) the knowledge and skills necessary to help individuals overcome functional limitations, (5) the willingness to become involved in this interpersonal process, and (6) the knowledge of oneself and one's own skills and limitations.

BELIEF THAT CLIENTS ARE UNIQUE INDIVIDUALS OF SIGNIFICANT VALUE. The belief that all human beings are worthwhile, valuable, and unique is an essential conviction for every counselor to have. This conviction must be present for you to relate to each client in a positive and constructive manner. This acceptance of and sincere belief in the client must be a felt experience and not an abstract philosophical concept. It does not mean that you must approve or disapprove of a particular act or like or dislike a particular trait manifested by a client; but rather, in spite of an act or regardless of a trait, you should have a genuine interest in the client and respect the client as an important, valuable, and worthwhile human being.

Furthermore, as a counselor you must understand that a client's self-perception and perception of the world constitute reality for that person. The beliefs, attitudes, feelings, and cognitive impressions that the client has about self and his or her environment strongly influence the way that person behaves. As a counselor you need to focus on understanding these perceptions, for they provide valuable vehicles for viewing the client's internal frame of reference, grasping the uniqueness of the client's world, and comprehending the meaning of the client's behavior.

Counselors who manifest this belief in their clients' sense of worth and uniqueness encourage their clients to develop a feeling of trust during the counseling process. This belief is communicated in a number of ways. Nonverbally, it is communicated by promptness, posture, and facial expressions; paraverbally, it is communicated by tonal quality; and verbally, it is communicated by responses that are sensitive to the feelings and attitudes of the client.

BELIEF THAT CLIENTS ARE CAPABLE OF CHANGE.   Your theoretical orientation and your basic assumptions about the nature of human beings will largely determine your belief regarding the kind of change and the amount or degree of change possible for any individual client. Counselors may hold distinct and varied opinions on the kinds of changes that they believe are possible, but all counselors must believe that individuals are capable of changing.

Counselors tend to be optimists, and they believe that all clients can, at least to some extent, modify their feelings, attitudes, cognitive structures, or overt behaviors. They recognize, moreover, that helping people to change is hard work and that sometimes they cannot help a particular person. When this occurs, the inability to help the person is not ascribed to the impossibility of change but rather to extenuating causes such as the client's unreadiness for change, the need to modify the client's environmental situation, or the counselor's lack of knowledge or inexperience in dealing with this client's particular type of problem.

As a counselor you must communicate your belief that your client is capable of changing. This communication should be accomplished by your actions and attitudes and not depend on verbal communication. The use of words such as "I believe you can solve your problem" will not, by and of itself, persuade your client of your belief. Other communication channels such as body gestures, facial expressions, and voice tone are more subtle but also more powerful ways of communicating your real attitudes and beliefs.

THE KNOWLEDGE OF HOW INDIVIDUALS FUNCTION.   Counselors need to understand the psychological principles that guide human behavior and to be aware of the environmental factors that influence this behavior. This knowledge is ordinarily acquired through advanced study in graduate psychology courses, which focus on understanding both *nomothetic,* or general, laws that are common to all human behavior and *idiographic phenomena,* or the unique ways that particular individuals behave under these general laws.

Knowledge of how individuals function is essential to the entire counseling process. It is important when you are trying to build trust for establishing a working relationship, attempting to explore and understand the factors that are delimiting the client's behaviors, deciding upon a particular treatment strategy, employing an intervention strategy in a sound and appropriate manner, and deciding when to terminate a case.

KNOWLEDGE OF HOW TO ASSIST INDIVIDUALS. Counseling requires more of you than a willingness to enter into a relationship, a belief in the value of individuals, a belief in change, and the knowledge of how people function. You must also possess important clinical skills that help you to assist individuals in finding the impediments that block their ability to undergo some changes and function at a more effective or higher level.

Clients are helped in many ways. What seems to work in one place or time may not work in another. A wide variety of approaches, methods, and theories of counseling are available. As a beginning counselor, however, you should avoid the meaningless smorgasbord approach that haphazardly takes a bit of this theory and a bit of that theory. As a counselor-in-training you should experiment, under supervision, with a variety of approaches that you can gradually integrate into a personal style through training and practice.

Learning how to assist individuals in overcoming their functional limitations is a time-consuming process of personal growth. Counselors-in-training normally obtain entry-level skills through didactic and experimental learning. They read and hear about counseling approaches. They try to emulate the behavior of expert counselors by shaping their responses to fit the pattern of the particular expert whom they are studying at that time. Through sequential clinical experiences such as simulations, role playing, supervised practice, and internships, you will learn to deal with conceptual and actual problems and thereby develop your own personal skills. Skill development is an ongoing process, and you should expect to continue to learn skills and add them to your repertoire throughout your entire professional career.

WILLINGNESS TO BECOME INVOLVED. Counselors must demonstrate their willingness to become involved in this interpersonal process called counseling. This commitment to share oneself goes beyond merely giving the time and energy required to assist another person. It includes the effort to bring as much of yourself as necessary into the helping relationship, and the ability to communicate to the client that nothing is more important at that moment than the client and what he or she has to say. Furthermore, this commitment requires you to be willing to concentrate fully on the client's internal frame of reference, to reach out to help a client understand himself or herself, to assist a client in understanding the change process and any impediments that may be present, and to take some risks in using yourself as an instrument to facilitate this change.

Counselors who have good feelings of their own self-worth, adequacy, and self-discipline transcend their own limitations and are free to give the necessary attention to their clients and to focus on ways to assist them. You can communicate this willingness to become involved by being warm, understanding, and sincere; by concentrating on the messages your clients are trying to communicate; and by giving genuine, unhurried, and sincere responses.

KNOWLEDGE OF ONESELF.   Counselors must have positive self-concepts and feel secure about themselves. You should be aware of your own feelings, attitudes, values, and motivations for working with others. You will need to know your own skills and to willingly acknowledge your limitations. You should be open to self-improvement and growth through additional learning and experience, and you need to acknowledge that all persons, including yourself, have a range of talents and limitations.

Counselors realize that they cannot help everyone with every functional limitation, and they must be able to recognize when a particular client's concerns require some specialized knowledge or area of expertise they may not possess. You will need to become sensitive to those areas and exercise appropriate discretion and judgment in referring the client to other specialists.

This process of knowing yourself and facing up to your own limitations has two important influences on counseling. First, the better you can understand and appreciate your own feelings, thoughts, and behaviors, the better you can understand and appreciate the feelings, thoughts, and behaviors of others. Second, counselors who are comfortable with themselves communicate an attitude of genuineness to the client. The client, sensing that genuineness, develops or confirms a sense of trust in the couseling relationship and thus will unfold more deeply his or her internal frame of reference and thus move the counseling process along.

THE COUNSELOR'S PROFESSIONAL EDUCATION AND TRAINING

The professional counselor is a person who has completed a graduate-level program of studies in an accredited institution of higher education and earned a master's or doctoral degree in one of the specialties in the field. Master's degree programs typically require students to develop a specialization in one of the following areas:

Counseling in Community Agencies
Marriage and Family Counseling
Mental Health Counseling
Rehabilitation Counseling
School Counseling
Student Affairs Practice in Higher Education

Doctoral degree programs are available in

Counselor Education and Supervision
Rehabilitation Counseling
Counseling Psychology

Standards for the education and training of counselors have been published by different professional groups (see Table 1-1).

Until recently, most counselors obtained their initial level of professional training in a one-and-one-half-year or 30-to-36-semester-hour master's degree training program. Although each of the major professional associations has promulgated its own training standards, these standards typically specify that counselors-in-training should complete a two-year or 48-to-60-semester-hour program to meet minimum professional requirements. These standards normally outline a common core of knowledge and some unique competencies for the specialization areas. For example, CACREP standards indicate that students need to develop competencies in each of the following eight areas:

- human growth and development;
- social and cultural foundations;
- helping relationships;
- group methods;
- career development;
- appraisal;
- research and evaluation; and
- professional orientation.

Additional competencies are specified for each of these specialization areas.

Many counselors have become National Certified Counselors (NCC) by meeting the standards set by the National Board for Certified Counselors (NBCC), the certification affiliate of the American Association for Counseling and Development (AACD). This certification process was started in 1982 and provides a

TABLE 1-1
Education and Training Standards for Counselors

American Association for Counseling and Development (AACD): *Council for Accreditation of Counseling and Related Educational Progams* (CACREP)
    Doctoral programs in Counselor Education and Supervision
    Marriage and Family Counseling
    Mental Health Counseling
    School Counseling
    Student Affairs Practice in Higher Education

American Association of Marriage and Family Therapists (AAMFT)
    Marriage and Family Counseling

American Psychological Association (APA): *Counseling Psychology Division*
    Doctoral programs in Counseling Psychology

Council on Rehabilitation Education (CORE)
    Rehabilitation Counseling

---

national registry of certified counselors. To become nationally certified you have to meet certain specified educational standards, obtain appropriate professional counseling experience, and successfully complete a written national examination. This examination is designed to test your knowledge of the eight content areas specified in the CACREP standards (see page 7). Certified counselors must show evidence of continued professional growth and development to maintain this credential. NBCC also offers counselors the opportunity to obtain specialty certification in school counseling, career counseling, and gerontological counseling.

Counselors can also be nationally certified in two other areas. The American Mental Health Counselors Association (AMHCA) has established requirements for obtaining special recognition as a Certified Clinical Mental Health Counselor, and the Commission on Rehabilitation Counselor Certification (CRCC) has developed criteria for becoming a Certified Rehabilitation Counselor.

Professional counselors may also obtain a professional license in the majority of states in the United States (34 in 1990). The movement toward licensing has been actively pursued by AACD and the various state associations for counseling and development. Because each state controls its own professional credentialing legislation, wide variations exist in the terminology and functions defined in each of these statutes.

Professional counselors keep up their education and training by reading professional journals, by attending workshops and various training programs, and by actively participating in one of the associations devoted to counseling. The largest professional association for counselors is the American Association for Counseling and Development (AACD). This association has 16 divisions and organized affiliates devoted to different aspects of counseling (see Table 1-2). AACD and its divisions conduct an annual convention, and each division publishes a newsletter and a journal of particular interest to the members of that division.

Professional counselors often work very closely with other helping professionals such as psychologists and social workers. Standards for the education and training of members of these allied professions have been determined by their respective professional associations, the American Psychological Association (APA) and the National Association for Social Workers (NASW).

Ethical Considerations for the Counselor

The overall goal of the counseling process is to help the client become a more effectively functioning person. In this process the counselor's role is to assist the client in modifying his or her behavior or in choosing between alternative courses of action, or perhaps to support the client while he or she is experiencing some trauma in the course of life. Ultimately, of course, clients are responsible for choosing whatever behavior they want to adopt, what direction they wish to go, and how they wish to resolve the traumatic episodes in their lives. The counselor's primary job is assisting the client; nevertheless, the counseling

TABLE 1-2
Divisions and Organizational Affiliates of AACD

American College Personnel Association (ACPA)
American Mental Health Counselors Association (AMHCA)
American Rehabilitation Counseling Association (ARCA)
American School Counseling Association (ASCA)
Association for Adult Development and Aging (AADA)
Association for Counselor Education and Supervision (ACES)
Association for Humanistic Education and Development (AHEAD)
Association for Measurement and Evaluation in Counseling and Development (AMECD)
Association for Multicultural Counseling and Development (AMCD)
Association for Religious and Value Issues in Counseling (ARVIC)
Association for Specialists in Group Work (ASGW)
International Association of Addictions and Offenders Counselors (IAAOC)
International Association of Marriage and Family Counselors (IAMFC)
Military Educators and Counselors Association (MECA)
National Career Development Association (NCDA)
National Employment Counselors Association (NECA)

relationship has a strong bearing on the counseling outcome. In the relationship, the counselor is the helper, the nurturer, the more effective person, or the mentor and, hence, exerts considerable influence on the client. The counselor can alter counseling outcomes in significant ways.

CODE OF ETHICS. Clearly, counselors have an obligation to work for their client's best interest. Toward this end, experienced counselors as well as counselors-in-training should be knowledgeable about the standards and principles of conduct that have emerged over the years through the consensus of experienced counselors. These ethical standards or principles are codified and promulgated by national professional associations such as the American Psychological Association (APA), the American Association for Counseling and Development (AACD), and the American Association of Marriage and Family Therapists (AAMFT). They provide guidelines for appropriate ethical behavior in dealing with issues relating to professional practice. These written codes of ethics fulfill several functions: first, they help clarify the counselor's responsibility to the client; second, they provide some guidelines for resolving conflicts that may arise in a particular situation; and third, the codes provide a standard by which members of the profession may be judged.

Obviously, ethical standards, by their very nature, offer a discrete number of statements about general professional behavior. Because of the discreteness and the generality of these statements, they leave gray areas and offer limited directions for some issues that may arise in professional practice. You must use your own judgment in applying these standards to specific situations. Six divisions of AACD (ACPA, AMHCA, ARCA, ASCA, ASGW, and NCDA) have published their own codes for dealing with issues of primary concern to members of these divisions, and the National Board for Certified Counselors (NBCC) has published its own code of ethics to guide nationally certified counselors. As a counselor-in-training, you should familiarize yourself with the ethical standards developed by professional associations such as AACD, AAMFT, and APA, and the divisions of AACD that you are interested in joining. You should review several other publications concerning ethical issues such as

Specialty Guidelines for the Delivery of Services by Counseling Psychologists (APA, 1981);

Ethical Standards Casebook (Herlihy & Golden, 1989);

Issues and Ethics in the Helping Professions (Corey, Corey, & Callanan, 1988);

Ethical, Legal, and Professional Issues in the Practice of Marriage and Family Therapy (Huber & Baruth, 1987);

Ethics and Values in Psychotherapy (Rosenbaum, 1982);

Ethical and Legal Issues in Counseling and Psychotherapy (Van Hoose & Kottler, 1985).

IMPORTANT ETHICAL ISSUES IN COUNSELING

Important ethical concerns revolve around four major issues: the client's welfare, the competency of the counselor, the client's right to privacy, and the value issues inherent in the counseling process.

The Client's Welfare

The primary responsibility of every counselor in the counseling relationship is the welfare of the client. Clearly, counselors do not work in a vacuum; they also have responsibilities to themselves, to the agencies and institutions for whom they work, to their profession, and to society in general. Putting the welfare of the client first means that your top priority is helping the client and ensuring that the client obtains appropriate assistance from either yourself or another helping professional. If you are involved with the client in another role such as teacher or administrative superior, or in a close social relationship, the counseling relationship will be confounded and your ability to be objective will be hindered. Consequently, it is generally better to avoid counseling individuals with whom you have another relationship. Refer these cases to another counselor.

The Competency of the Counselor

Counselors must present themselves and their qualifications in honest, truthful ways and work within their own areas of professional expertise. As a counselor trainee who has had little counseling experience in which to develop expertise, you should present yourself as a counselor-in-training and work within the parameters for which you are obtaining adequate supervision. When you finish your formal training, you should apply for the positions you are qualified for, and you should obtain the credentials and licenses these positions require. When working with clients, you should employ intervention strategies that you are competent in using and that are appropriate for the situation. If a client has concerns that you cannot handle, a referral to an appropriate professional or agency should be made. To ensure that you maintain your competency as a counselor, you should keep up with developments in the field by reading and attending workshops and professional meetings, and through continued supervision.

The Client's Right to Privacy

Individuals have a right to the privacy of their own lives. When clients discuss their concerns with counselors in the counseling process, they reveal considerable information about themselves. In this sharing process, the client places a certain amount of trust and faith in you, and you assume an obligation to protect the client's right to privacy. This protection involves two major concepts that beginning counselors need to be aware of: confidentiality and informed consent, and recording the interview.

CONFIDENTIALITY AND INFORMED CONSENT. To the beginning counselor trainee, confidentiality often implies that the counselor should never share any information about the client with any other person. This rigid interpretation is too all-encompassing and, indeed, may be counterproductive.

Information about the client should be used for the welfare of the client. When you honestly and professionally believe that information about the client should be shared with others, you should be prepared to do so. However, you will need to obtain the client's consent, and make the client aware of the material to be shared and under what circumstances and with whom it will be shared.

Counselors obtain information about clients from the clients themselves during the counseling process and from other sources. In other words, this information can be internally (interview) or externally (noninterview) based. Furthermore, counselors interact with professional supervisors and colleagues, professional workers in other disciplines, and other significant persons in the life of the client, all of whom are interested in the client's welfare. Generally speaking, when interacting with significant other persons, counselors discuss information about their clients in the following hierarchical order:

1. Interview and noninterview data are frequently shared with professional supervisors and colleagues.
2. Noninterview data are often shared with professionals in other disciplines and sometimes with significant other persons.
3. Interview data are not shared with others unless there is a compelling reason for doing so.

Counselors and counselors-in-training share interview data with their supervisors and colleagues to benefit from the experiences of others and thus develop their skills and be better prepared to help their clients. In fact, most counselors-in-training are required to share their notes and audio- or videotapes of their counseling sessions with their supervisors and colleagues for critical reviews. Counselors share noninterview data and inferences from interview data with professionals in other disciplines and with significant others to gain their cooperation and expertise in assisting the client or to keep them informed of the progress of the case. Normally, the counselor does not share interview data with professionals in other disciplines and significant others unless clear danger to human life exists or unless the client specifically consents to divulging the information.

Clients are usually not bothered when they are told that interview and noninterview data will be shared with professional supervisors and colleagues. They are pleased to know that you are in contact with other skilled professionals and that there are many resources available. Likewise, they usually have no problem when you share noninterview data and inferences from interviews with professionals in other disciplines and significant other persons. They may or may not object to your sharing interview data with individuals from these two groups. If clients are aware of the reasons why you would like to share this information, objections normally vanish.

Two general principles that must be kept in mind when sharing both interview and noninterview data with others are (1) obtain the client's informed consent, and (2) reveal everything that is essential and absolutely nothing that is not essential.

RECORDING THE INTERVIEW.   Recording the interview in one way or another is essential for several reasons: to remind yourself what happened with a particular client in a particular session, to review your work with your supervisor or professional colleagues, and to show your own growth and development as a counselor.

There are several ways to record a session: audiotaping, videotaping, and note taking. Audiotaping and videotaping are easy and fruitful ways for a beginning counselor trainee to record counseling sessions because of the simplicity of audiocassette tape machines and the portability of video equipment.

These methods provide an excellent means for your supervisor to monitor your growth over time and to see how you handle a given situation. The supervisor can review your moment-to-moment interaction with the client and give feedback, pointing out what was well done, what was not so well done, and what was left undone. It is imperative, of course, that the equipment not get in the way of the counseling relationship. Beginning counselor trainees often need several experiences with recorders so that they can (1) feel comfortable working with the machinery and hearing their own voices and seeing themselves; (2) gain some expertise in obtaining the informed consent of their clients in truthful but nonanxiety-provoking ways; and (3) learn how to place recorders, microphones, and cameras in inconspicuous places so the client and counselor do not have their attention drawn to these devices and, hence, will focus on the counseling process and not on the recording process.

More experienced counselors and counselors who work in settings where sessions cannot be taped may rely on written notes taken either during or immediately after the counseling session.

There is not one specific or correct way to take these notes. As a counselor trainee, you should follow the recommendations of your immediate supervisor until you develop your own style and preferred method. Generally, notes taken during the interview record factual information stated, ideas and feelings mentioned, and plans or goals discussed with the client. Notes written immediately after the session contain a résumé of the points discussed, a summary of the dynamics of the session, and a statement about tentative goals for the next meeting.

Neophyte counselors who are trying to develop many skills simultaneously must be cautioned against the two extreme positions: compulsion and scantiness. At the one extreme, trainees may be so intent on taking notes that either they do not concentrate on the relationship and use the note taking as an excuse or a refuge from developing their communication skills or they become interrogators rather than counselors. At the other extreme, trainees may be so intent on the session or concerned with their skills that they fail to record important and significant events.

When written notes are used to record interviews, clients should be aware that these notes are being kept. When taken during the session, notes should be made in such a way as to be as inconspicuous as possible. They should be brief enough so that the counselor/client relationship is not affected and sufficiently long to record the significant elements.

When recording a session, be honest with your client. Explain the purpose of the note taking or the audio- or videotaping and explain with whom you expect to share the case materials. Do not write anything you do not want your interviewee to see, and allow your client to read your notes or listen to or see the recording if he or she wishes to do so. Remember, counseling is a collaborative effort, and any notes or records of the sessions are done for the client's welfare.

Notes and recordings of the interview need to be kept in a secure place to which only the counselor has immediate access. And normally they should be shared only with those who are helping the counselor—namely, his or her supervisor or professional colleagues.

Value Issues Inherent in the Counseling Process

Counseling is not a value-free human endeavor. All counseling is intimately involved with cultural, moral, and ethical values related to the three major spheres of life: the educational/vocational dimension, the marital and family dimension, and the social/cultural dimension. Both counselors and clients bring to the counseling relationship deeply cherished values concerning education, work, marriage and family issues, and the individual's obligations and responsibilities to those in his or her immediate environment as well as those incumbent upon him or her as a citizen.

Neither clients nor counselors leave their values at the door to the counseling office. The only value that may appear overtly in a counseling session is the dignity and respect that both participants reveal in their treatment of one another. However, other values are usually implicit in the relationship and are not obvious, principally because both the client and the counselor are working under the same value system and do not need to discuss them. Some examples include:

- the school counselor advising the college preparatory student about various colleges—both have implicitly accepted the value that "a good education is most desirable";
- the vocational counselor educating an unemployed head of household about job placement strategies—both have implicitly accepted the "Protestant work ethic"; and
- the behavioral counselor treating the obese client—both have accepted the value that being overweight is unhealthy.

Generally speaking, value issues become critical in the counseling process when one of the following situations occurs.

- The values of the client and the counselor are different.
- The values of the client are causing some difficulty in his or her environment.
- The counselor would like to employ a treatment that he or she believes, in the long run, will help the client, but knows that the treatment has some side effects, involves some pain or other harm to the client, or could lead to results that may leave the client worse off than if he or she did not receive the treatment.

When the client and the counselor have different values regarding an issue that is relevant to the counseling relationship, the counselor needs to remain aware of these differences and respect the client's right to his or her own values about a particular issue. The United States is a pluralistic society, and counselors must work within that system. Because the goal of any counseling relationship is to help the client resolve his or her own problem, you should, if at all possible, try to work within the client's frame of reference and value system to find a solution. However, sometimes you may find that it is impossible to do this. When this occurs, the conflict should be discussed openly with the client, and if further counseling proves impossible, a referral to another counselor is mandated.

When the client's value system is causing the client difficulty, the counseling is clearly value dominated. Again, the counselor must remember that the overall goal of the process is to help the client help himself or herself. Therefore, you need to help the client discuss his or her values in the client's own environment, and help the client resolve the difficulty or at least cope with the situation in a more effective way.

The third critical value situation arises when a proposed treatment strategy involves some risk to the client and may cause the client some physical or psychological harm or otherwise leave the client in a worse state than he or she was before treatment. Clearly, risk-taking is a normal human behavior. We take chances in virtually all things we do. Usually we do things that involve odds that are overwhelmingly in our favor, but we should encourage our clients to undertake a risky treatment when the following guidelines have been met:

- the client's present situation is very bad or is deteriorating, and some treatment involving risk is necessary;
- the client fully understands the risk and consents to the treatment; and
- the proposed treatment has been reviewed by an objective panel of professional supervisors or experts, and the panel agrees that the treatment is warranted.

LEGAL CONSIDERATIONS FOR THE COUNSELOR

Counselors should be aware of certain legal issues related to the practice of counseling. The major legal issues relate to confidentiality and privileged communication, specific federal and state legislation, and malpractice.

First, the distinction between the terms confidentiality and privileged communication must be understood. *Confidentiality* is an ethical term that refers to the client's right to privacy, guiding counselors to disclose information only with the informed consent of the client. Under certain unusual conditions counselors are legally bound to breach confidentiality. These situations involve clear and immediate danger to the client or to a third party—for example, child abuse, potential suicide, or serious harm to another person. *Privileged communication* is a legal term that refers to the right of the client to discuss matters with an appropriate person and not have this communication reported in a court of law. According to Herlihy and Sheeley (1987), professional counselors have been granted privileged communication rights in 14 states and school counselors have limited privileged communication provisions in 20 states.

Second, counselors should become familiar with both the national and local laws that guide their practice. At the national level the Family Educational Rights and Privacy Act (FERPA), commonly known as the Buckley amendment, stipulates that clients or their parents have the right to examine all official records. The Education of All Handicapped Children Act (Public Law 94-142) outlines the responsibilities of school counselors working with handicapped children. Because state laws govern what happens within state boundaries and these laws vary, counselors should become familiar with the laws of the state in which they work.

Third, counselors may be sued for malpractice when they fail to render the proper service either through ignorance or neglect, or when some injury or harm results from acting on the counselor's advice or recommendation. Fischer and Sorenson (1985), Hopkins and Anderson (1985), and Woody et al. (1984) are excellent books dealing with malpractice issues and other legal aspects of counseling and the helping professions. Counselors can obtain liability insurance through the American Association for Counseling and Development.

SUMMARY

This chapter has presented the counselor and the counseling process. After reading it you should be familiar with several important aspects of counseling. First, you should be able to discuss some of the important personal and professional characteristics that are important for counselors. Second, you should be able to describe the professional educational and training requirements for counselors. Third, you should be able to name the major professional counseling associations. Fourth, you should be able to identify the major ethical standards or codes of conduct that guide counselors. Fifth, you should be able to discuss the major ethical issues involved in the counseling profession. And finally, you should be able to identify some of the major legal issues that affect the counselor.

REFERENCES AND SUGGESTED READINGS

American Association for Counseling and Development. (1988). *Ethical standards* (rev. ed.). Alexandria, VA: Author.

American Association for Marriage and Family Therapists. (1985). *AAMFT code principles for marriage and family therapists.* Washington, DC: Author.

American Psychological Association. (1989). *Ethical principles of psychologists* (rev. ed.). Hyattsville, MD: Author.

American Psychological Association. (1981). *Specialty guidelines for the delivery of services by counseling psychologists.* Hyattsville, MD: Author.

Avila, D. L., & Combs, A. W. (1985). *Perspectives on helping relationships and the helping professions.* Boston: Allyn & Bacon.

Benjamin, A. (1987). *The helping interview* (4th ed.). Boston: Houghton Mifflin.

Boy, A. V., & Pine, G. J. (1982). *Client-centered counseling: A renewal.* Boston: Allyn & Bacon.

Brammer, L. M. (1988). *The helping relationship* (4th ed.). Englewood Cliffs, NJ: Prentice-Hall.

Corey, G., Corey, M., & Callanan, P. (1988). *Issues and ethics in the helping professions* (3rd ed.). Pacific Grove, CA: Brooks/Cole.

Fischer, L., & Sorenson, G. P. (1985). *School law for counselors, psychologists, and social workers.* New York: Longman.

Herlihy, B., & Golden, L. (1989). *Ethical standards casebook* (4th ed.). Alexandria, VA: AACD Press.

Herlihy, B., & Sheeley, V. (1987). Privileged communication in selected helping professions: A comparison among statutes. *Journal of Counseling and Development, 65,* 479–483.

Hopkins, B. R., & Anderson, B. S. (1985). *The counselor and the law.* Alexandria, VA: AACD Press.

Huber, C. H., & Baruth, L. G. (1987). *Ethical, legal, and professional issues in the practice of marriage and family therapy.* Columbus, OH: Merrill.

National Association of Social Workers. (1979). *Code of ethics.* Washington, DC: Author.

Rogers, C. R. (1983). *Freedom to learn for the eighties.* Columbus, OH: Merrill.

Rosenbaum, M. (Ed.). (1982). *Ethics and values in psychotherapy.* New York: Free Press.

Van Hoose, W. H., & Kottler, J. A. (1985). *Ethical and legal issues in counseling and psychotherapy* (2nd ed.). San Francisco: Jossey-Bass.

Woody, R. H., & Associates. (1984). *The law and the practice of human services.* San Francisco: Jossey-Bass.

2 | *The Effectively Functioning Person*

INTRODUCTION

This chapter presents a brief overview of several areas where clients frequently manifest some difficulties in their lives. Helping clients become more effective, more fully functioning, and more independent is the ultimate goal of any counseling relationship and is implicit in all counseling approaches. The sections of this chapter should help you understand many of the ways in which individuals who function quite effectively differ from those who function less effectively. After careful review of the material in this chapter you should be able to:

- describe how a person functions effectively or ineffectively in one or more of the following dimensions of life: need satisfaction, stress and the coping processes, developmental task attainment, social contact and interpersonal relationship skills, and other personal or characteristic attributes; and
- discuss some of the major problems that can impede the effective functioning of individuals, such as handicapping conditions, serious vocational problems, substance abuse problems, and psychological disorders.

IMPORTANT PSYCHOLOGICAL DIMENSIONS OF FUNCTIONING

Individuals who are functioning effectively usually (1) satisfy their needs in appropriate ways, (2) deal with the stresses of life and their emotional reactions by using effective coping processes, (3) learn tasks that are appropriate to their developmental stage, (4) have worthwhile social interactions and interpersonal relationships, and (5) demonstrate other positive attributes. Individuals who are functioning less than optimally often manifest problems in one or more of these areas.

Need Satisfaction

Needs, drives, and motives are the energizing forces that propel us toward or away from some action or thing that we require for our survival or well-being (London & Exner, 1975).

Many theorists postulate that all human beings have a primary motivation that provides the general direction or outlines the major thrust of human life (Maddi, 1989). Freud (1966) maintained that we all have two primary tendencies: first, to gratify our instinctual drives, and second, to minimize our feelings of guilt and punishment. Because these two basic tendencies have different aims, they impel us in different directions. Therefore, according to Freudian theory, we need to make compromises between these tendencies in order to function effectively.

The humanistic theoreticians such as Adler (1964), Maslow (1970), and Rogers (1959) have a different view of the primary thrust in our lives. According to these theoreticians, this primary force is a general striving or growth motivation to fulfill our inherent potential. Adler, Maslow, and Rogers held slightly different positions about the source and nature of this primary motivation but not about the essential belief that life has an overriding goal-directed purpose. The behavioral psychologists, such as Skinner (1953), maintain that we are not born with an overriding central purpose or thrust in life but rather that all our behaviors are a result of the learning process.

Regardless of these different positions about the existence or the source of a primary tendency in our lives, most authorities agree that certain needs are common to all of us. Needs are usually discussed in terms of two separate domains. Needs related to our physiological nature and required for our bodies to function effectively are referred to as *drives*. Needs associated with social functioning and important for our psychological well-being are referred to as *motives*.

In the physiological domain we have certain needs that must be met if we are to remain alive, comfortable, and in good health. These biological imperatives, drives, or survival needs include the need for food, liquids, clothing, shelter, activity, rest, and appropriate body temperature. An important part of this physiological need structure is the tendency for our bodies to seek *homeostasis,* or a balanced state of equilibrium or stability. When we are in a state of deprivation (in need of food or liquids) or in a state of discomfort (too hot or too tired), this drive for homeostasis causes us to change our activity to improve our body's well-being. Because these needs are activated by a lack of something that the body requires, they are considered to be deficiency motives by Maslow (1970) and primary drives by the behavioral psychologists (Goodstein & Lanyon, 1979).

In the psychological domain we also have certain needs that we must satisfy to function in an adequate and appropriate manner. As these needs primarily concern our relationship to society, they are considered by behavioral psychologists to be *socially derived needs* learned in the process of socialization (Goodstein & Lanyon, 1979). Those who hold this position refer to them as acquired needs or secondary drives. However, other experts, such as the Adlerians

(Dreikurs & Soltz, 1964), the humanistic psychologists (Maslow, 1970), and the Transactional Analysts (Elson, 1979) imply that certain of these needs are so imperative for our well-being that they are neither acquired nor secondary in nature. Regardless of the theoretical differences concerning the origin of these psychological needs, they do motivate much of human behavior and must be satisfied for us to function effectively.

Because our psychological needs greatly influence how we function, counselors must have a knowledge of these needs and how they control our behavior. The psychological needs that counseling and psychotherapy authorities consider extremely important are summarized in the following paragraphs.

THE NEED FOR SAFETY, SECURITY, AND SELF-PRESERVATION. According to Maslow (1970), all human beings need to feel safe and secure, and to live in a nonthreatening environment. We are motivated to shun painful and dangerous situations, avoid chaos in our lives, minimize anxiety, and protect our sense of self. This self-protection need can cause any one of us to become quite assertive or aggressive. According to Dinkmeyer, Dinkmeyer, and Sperry (1987) and Sweeny (1989), this need impels discouraged children, who mistakenly feel attacked and deeply hurt, to seek revenge in order to protect themselves from further challenges to their safety and security.

THE NEED FOR STRUCTURE AND ORDER. In addition to feeling safe and secure, we all need to have structure and order in our lives (Maslow, 1970). Elson (1979) and Wollams and Brown (1979) describe how we structure our time in different ways. This can range from becoming involved in very meaningful activities or relationships to engaging in quite meaningless ones. Individuals who are functioning effectively typically have better structure in their lives than those who are functioning less effectively.

THE NEED FOR ATTENTION, CONTACT, AND POSITIVE REGARD. We are all social entities, and we have a basic need to obtain attention and establish contact with other human beings that impels us to make physical and emotional contact with others. The need for recognition or strokes is extremely important and powerful. Those of us who do not have these needs satisfied in a positive manner by receiving smiles, hugs, or words of approval will seek to have them satisfied in negative ways by seeking frowns, slaps, or words of admonition. According to Rogers (1972), satisfying this need for positive regard is crucial to the healthy personality. Dinkmeyer, Dinkmeyer, and Sperry (1987) maintain that children who misbehave are very often seeking attention but are doing so in ill-advised and inappropriate ways.

THE NEED FOR AFFECTION, BELONGINGNESS, AND LOVE. Closely aligned to the need for attention is the need to engage in reciprocal relationships with significant others. Every individual strives to find and maintain a relationship with other individuals and to be a member of some group. This need to belong, to love and be loved, and to relate to other human beings in significant ways influences much of human behavior. When this need is not met it is one of the more prominent causes of behavioral problems among individuals (Maslow, 1970).

THE NEED TO FEEL UNIQUE AND FOR POSITIVE SELF-REGARD.   Each one of us needs to feel unique and worthwhile and to have a positive feeling of self-worth. Having a positive self-regard is crucial for being well-adjusted or fully functioning (Rogers, 1972). Furthermore, out of this need to feel unique, we all need to take time occasionally to reflect, meditate, and be by ourselves. Extremely discouraged children express this need by withdrawing from situations and giving up on some or all aspects of their behavior (Dreikurs & Soltz, 1964). This withdrawal can be manifest in a variety of ways such as being sick, failing course work, or being in a sad or depressed state.

THE NEED TO INFLUENCE AND CONTROL ONE'S ENVIRONMENT.   We all have the need to influence and control our environment for our own benefit. This need impels us to learn things and engage in activities that are important to us. People sometimes seek to fulfill this need to achieve and gain control in ways that are ill-advised. Children, in particular, can assert this need for power and control of their environment in ways and at times that are most inappropriate (Sweeny, 1989). Parents, teachers, and significant other adults ought to be aware of this and take suitable steps so that power struggles can be avoided.

Needs are complex energizing processes that influence the direction and the intensity of our behavior. The strength of any particular need at any particular time differs among us and also within us. Whether a particular need or motive will activate any of us toward a particular behavior is the product of several factors:

- The opportunity for the need to be satisfied must be available; circumstances beyond our control can prevent a need from being actively pursued.
- The appropriate target, object, or person to whom our need is directed must be present.
- The notion of having at least a reasonable chance for success must be perceived.
- Our anxiety level should not be too intense; being overly anxious frequently leads to self-defeating behaviors.
- The incentive value of other needs must be less intense.

In the process of helping clients, it is important for counselors to understand how these needs and motives influence behavior. Clients who experience frustrations or manifest ill-advised and maladaptive behaviors often have unsatisfied needs, or they may be attempting to meet their needs in inappropriate ways. As counselors, we need to help our clients explore and understand these factors and aid them in developing more effective strategies for meeting their important needs.

Stress and the Coping Process

STRESS.   At every stage of life it is normal for any one of us to encounter challenging situations or taxing and potentially stressful conditions. The major sources of psychological stress are frustrations, conflicts, changes, and pressures (Weiten, Lloyd, & Lashley, 1991). These challenging conditions often occur at transitional points in our lives, and they always produce anxiety or some other emotive reactions.

When we are *frustrated,* either an external or an internal obstacle prevents us from acting or obtaining some desired goal. External blockages can be caused by social, economic, legal, or other environmental conditions. Internal blockages are often related to some personal limitation such as a handicap, insufficient knowledge, or lack of a specific ability, skill, or talent. Some frustrations are quite brief and relatively insignificant, and although stressful, the blockage can be removed in a short time. For example, a flat tire may prevent you from keeping an important appointment, but the tire can be changed and the appointment rescheduled. Other frustrations are much more serious—for example, the parents of an addict who are unable to involve their child in a suitable treatment program are probably quite frustrated and may be experiencing chronic stress.

When we are faced with a *conflict* we have to choose between two or more incompatible alternative courses of action. There are three classic types of conflicts. In the *approach/approach* conflict you must choose between two or more incompatible goals or situations all of which are equally desirable; for example, people in a restaurant who are presented with a list of desirable entrées or desserts. Normally, this type of conflict is the least stressful since no matter what one chooses the outcome is pleasant. In the *avoidance/avoidance* conflict you must select from two or more undesired objects or events that you want to avoid. This conflict might be experienced by the student who does not want to study a foreign language but must take one to graduate. Typically this type of conflict is very stressful and frequently leads to delaying the decision in the hope that a better alternative might develop. In the *approach/avoidance* conflict you want to obtain an objective that has both positive and negative aspects—such as when you want to buy something very much but don't want to spend that amount of money. Conflicts are often a result of more complex situations and can involve a matrix of goals that you desire and situations that you wish to avoid (Weiner, 1974).

Any *change* in your life, whether perceived as positive or negative, may also cause stress. In a mobile society such as ours, some people have experienced considerable shifting in their lives. These changes, which may have been in educational or occupational endeavors, financial conditions, or geographical locations, have exposed individuals to a wide variety of new or different demands and increased the potential for stress.

A fourth cause of stress is *pressure.* According to Weiten, Lloyd, and Lashley (1991) this pressure is caused by the expectations and demands for us to act in ways deemed appropriate by society, our reference groups, or ourselves. This pressure may be the expectation for us to *perform* or achieve at a certain level in some area—for example, a student may be expected to earn honors in school work. In other cases, the pressure may be to *conform*—a young person may be expected to like a certain type of food, music, or recreational activity simply because her or his peers like that food, music, or recreational activity.

EMOTIONAL REACTIONS.   Anger, anxiety, and sadness are three major emotional reactions we experience when faced with a frustration, a conflict, or another stressful condition. Each of these feelings can vary in intensity from a mild reaction to an extremely strong one. Anger, our typical reaction to a frustration, can range from being slightly annoyed to being in a rage. Anxiety, which we normally experience when we face a situation that has an uncertain outcome, can vary from a minor apprehension to an immobilizing fear. And sadness, which may be our reaction to a bad experience, can range from feeling glum or slightly dejected to sorrow or grief or to deep depression (Weiten, Lloyd, & Lashley, 1991).

COPING STRATEGIES.   The active efforts that we employ to solve our concerns, ameliorate the stressful conditions we experience, and lower our anxiety are our coping strategies (Kagan & Havemann, 1988). Some of the ways that we cope are proactive, whereas others are more reactive. When we are functioning effectively, we handle stressful concerns and any thwarting conditions we encounter in productive ways by using a wide variety of adaptive, healthy, and constructive coping skills or positive adjustment mechanisms. When we are functioning less effectively, we typically use more maladaptive, unhealthy, and counterproductive coping skills or defense mechanisms (Bernard & Huckins, 1976).

Proactive or constructive coping strategies involve taking direct and effective steps to handle a given issue, problem, or demand of life (Coleman, 1979; Nikelly, 1977; Torrance, 1965). Employing positive coping strategies does not automatically ensure a successful outcome in any particular situation. However, since the coping process is an ongoing one that is repeated over and over again, the use of these constructive actions generally leads to positive outcomes. These proactive skills have been classified into various types by Moos and Billings (1982) and Weiten, Lloyd, and Lashley (1991) according to their focus or goal. These coping strategies involve efforts to:

> *Appraise the situation in realistic ways.* This involves making an accurate and realistic evaluation of the situations and circumstances that cause the stressful events in our lives. This process should provide a clear understanding of these problems and their ramifications. Having a clear perception of a problem and the circumstances that caused it is an important coping process of and by itself.
>
> *Use appropriate problem-solving skills.* Learning how to solve, modify, or circumvent the problems you face in life is a major coping process. This includes mastering the steps necessary to choose wisely among

alternative courses of action, learning how to accept the compromises that have to be made in your life, accepting and living constructively with a blockage that cannot be removed, acting assertively to remove those conditions that are preventing you from taking appropriate actions, and when necessary, learning new skills to enhance your competence.

Deal effectively with one's emotional reactions to stress. It is often important to reduce the emotional reactions caused by stress. This can be accomplished indirectly by either of the methods mentioned above or directly by learning how to release your pent-up emotions in mature and socially acceptable ways, learning how to put your problems aside for a while, and learning effective relaxation techniques.

Maintain one's body in good physical condition. Keeping our bodies in reasonably good physical condition enables us to deal more effectively with the demands placed on us by any stressful condition. Learning to eat nutritional foods, developing and maintaining good hygienic habits, and engaging in a reasonable amount of exercise are activities that can not only help prevent but can also serve to ameliorate the problems of stress.

Nonconstructive coping skills tend to be reactive rather than proactive. Although these tactics can relieve stress or any other emotional reaction temporarily, they ultimately have serious negative physical and psychological repercussions. These maladaptive tactics can be grouped into the following general patterns or styles: aggressive, defensive or transformed, self-blaming, and withdrawn.

Aggressive coping reactions occur when we attack another person verbally or physically. This type of reaction is often associated with anger. Anger can take various forms such as directing our feelings at the person or thing that caused the stress or problem, venting our hostility at an innocent party and thus finding a scapegoat, or having a generalized raging anger and acting in an aggressive manner with everyone.

Defensive or transformed coping reactions are present when we change or disguise the cause of our frustration or thwarting condition. These reactions typically take the form of one of the classic defense mechanisms: compensation, projection, rationalization, reaction formation, or sublimation.

Self-blaming coping reactions are manifest when we engage in highly critical negative self-talk that assumes we are responsible for the problems in our lives and blaming ourselves for all our misfortunes.

Withdrawn coping reactions are evident when we either passively accept or inappropriately remove ourselves from dealing with the frustrations, conflicts, and other concerns of life. This type of reaction is frequently associated with an emotional state of sorrow and depression. This retreat from the challenges of life may be exhibited by fantasizing, regressing, mental wandering, or apathy. An apathetic reaction is probably the most maladaptive form of the reactive or poor coping mechanisms.

Clients who manifest poor skills in coping often have weak convictions about their abilities to overcome the stressful conditions they encounter. Consequently, any proactive attempts they make are rather limited in methodology, duration, and energy expended. Counselors who have clients with ineffective coping skills frequently need to develop treatment strategies that assist clients in improving their feelings of self-worth, appraising their situations more realistically, and marshaling their resources more efficiently. Normally, such treatment strategies employ a graduated or step-by-step approach and incorporate one or more of the interventions outlined in Part 2.

Developmental Task Attainment

As we progress through life from infancy to old age, certain physiological and psychological changes occur within us. These changes and the theories or models that attempt to find the broad underlying principles that describe these changes are investigated in the field of developmental psychology. Numerous theories try to explain how we change over our life span. Some theories focus on one aspect of behavior such as cognitive development (Piaget, 1966) or moral behavior (Kohlberg & Kramer, 1969), whereas other theories analyze one period of life such as childhood, adolescence, middle age, or maturity and old age (Biehler, 1976; Conger, 1983). This section of the text outlines some of the well-known theories of development, cites some references that you may wish to review, and makes suggestions about working with clients who have not mastered the appropriate developmental tasks.

Developmental psychologists frequently divide our life span into a series of *stages* or dynamic transitional periods in which all aspects of our physical and psychological development occur. Psychologists often differ on the exact meaning of the term *developmental stage* and the concepts and formulations they have about these stages (Lerner, 1986). Many theoreticians believe that life stages are structural entities that are universal (applicable to all of us), invariant (we must progress through these steps in a particular sequential order), and qualitatively rather than quantitatively different (distinctive laws govern us or unique phenomena apply to us at each stage). Furthermore, some theoreticians maintain that these life stages are transitional, overlapping, and continuous periods of time because elements of any stage may be present at another stage. The way we ascertain a person's developmental stage is by observing the modal or most common behaviors he or she manifests. Examples of stage theories include Freud's psychoanalytic theory of development, Piaget's theory of cognitive development, Kohlberg's theory of moral development, and Erikson's theory of psychosocial development.

Other psychologists either do not use the word *stage* to discuss developmental concerns or use this word to refer to a particular *phase* or period of life without implying all the characteristics of stages as outlined above. For example, Havighurst (1972) outlined certain developmental tasks or skills that we are expected to master at particular ages in our society, and Neff (1985) discussed how important it is for us to master certain work-related developmental tasks in order to become a contributing member of society. Clearly the age at which these developmental tasks are expected to be mastered can and does vary among and within different societies. Therefore, the age at which developmental tasks are appropriate is related to the expectations of a particular society or a unique reference group within that society, not related universally to a particular age group.

The major concepts of the developmental theories of Piaget, Erikson, and Havighurst, which are of interest to counselors, are briefly discussed in the following paragraphs.

Piaget's theory (1966) concerns the cognitive development of children and adolescents. According to his theory, our abilities to conceptualize objects in the world differentially occur at specific age-related stages. Our ability to accomplish the tasks of each stage depends on our biological readiness and how well we accomplished the tasks of the previous stage. Counselors who work with children and adolescents should be familiar with this theory and its possible relationship to a client's learning process. The four major stages of cognitive development, their approximate ages, and the major cognitive activities of each stage are outlined briefly in Table 2-1.

Erikson's theory postulates that crucial aspects of our psychological development occur during eight stages of life that each one of us is expected to progress through in our lifetime (Erikson, 1963). During each age-related period, we must learn how to deal effectively with specific issues or *psychosocial*

TABLE 2-1
Piaget's Stages of Cognitive Development

Sensorimotor period *(birth to 2 years)*	Knowledge of the world obtained through senses and motor activities. Development of cognitive patterns begins with immediate here-and-now experiences and then progresses to tendency to repeat interesting events, to an awareness that objects have a permanence, and to a curiosity about novel objects and events.
Preoperational period *(2 to 7 years)*	Gradually develop the ability to use symbols to represent objects; tend to focus on only one characteristic of an event or object at a time and ignore other features of the object; not able to think about a process or idea in an opposite or reverse way.
Concrete operational period *(7 to 11 years)*	Develop ability to think logically, to classify objects, and to understand how various items are interrelated; develop ability to reverse our thinking, which allows us to learn the fundamental arithmetic operations of addition, subtraction, multiplication, and division.
Formal operations period *(11 to 16 years)*	Develop the ability to think abstractly, to formulate hypotheses, to generate alternative solutions to problems, and to think about the future.

crises. Each crisis is a crucial event that leads to an outcome that lies somewhere between two bipolar opposites. The most successful and healthy outcome is at one end of the continuum; the least successful and most unhealthy outcome lies at the other end. Accomplishing this task successfully helps us gain the strength to cope with the crisis that occurs at the next stage. Individuals who go through the developmental stage with a negative or unsuccessful outcome will be limited in their future development.

Erikson's model suggests that counselors should be able to assess whether a client has successfully accomplished certain of these developmental challenges. However, Erikson's theory must be used with some caution. Hershenson and Power (1987) have pointed out that Erikson's theory is culturally bound to some extent and that each bipolar trait construction has built-in limitations. For example, the theory implies that failure to achieve autonomy by age three leads to shame and doubt. Clearly, autonomy is never fully accomplished by age three, nor is shame and doubt caused only by failure to achieve autonomy. Each of Erikson's life stages and their approximate ages, and the psychological crisis associated with each stage, are listed in Table 2-2.

Havighurst (1972) has a more detailed model of the developmental process than Erikson. Havighurst divides our life span into several major age-related periods and identifies certain developmental tasks that we should master during each of these specific periods. When we successfully accomplish these tasks, we obtain positive feedback from others and a deep sense of personal satisfaction and are well-prepared to meet the challenge of our next developmental period. If we do not learn the task at the appropriate time, it is more difficult to learn the task later in life, and we are poorly prepared to meet the tasks associated with the next developmental period.

Havighurst's list of developmental tasks can be used by counselors to assess a client's developmental progress. The specific age at which these tasks ought to be accomplished varies among cultures and among social classes within cultures,

T A B L E 2-2
Erikson's Life Stages

Life Period	Psychosocial Crisis	Desired Outcome
First year	Basic trust vs. mistrust	Acceptance, trust, and hope
Second year	Autonomy vs. shame and doubt	Control
Third through fifth year	Initiative vs. guilt	Purpose or goal
Sixth year to puberty	Industry vs. inferiority	Competence
Adolescence	Identity vs. confusion	Self-identity
Young adulthood	Intimacy vs. isolation	Mutuality and sharing
Middle adulthood	Generativity vs. self-absorption	Helping others
Mature age	Integrity vs. despair	Satisfaction with life

so these tasks must be seen as normative concepts rather than as absolute standards. The approximate age-related periods and the major developmental tasks associated with each of these periods are briefly outlined in Table 2-3.

Although the term "stage development" has different meanings, and not all experts agree on a particular theory of development, there is a general consensus that we all pass through certain age-related periods in which specific challenges should be met. Individuals who are functioning effectively accomplish the developmental tasks that are appropriate for their age and cultural group. They encounter a minimal amount of difficulty in progressing from one stage or phase of life to another. Individuals who manifest developmental difficulties have not mastered the educational, vocational, and social tasks that their society expects persons of that age group to have mastered.

During the early phases of the counseling process, you should assess the client's developmental progress. If the problem is related to a deficiency in developmental task attainment, a treatment strategy that focuses on developing self-understanding and awareness of one's potential and a methodology to overcome this deficiency needs to be implemented. Even though a wide variety of intervention strategies may be used, clients often benefit from an approach that incorporates the need to develop these tasks.

T A B L E 2-3
Havighurst's Developmental Stages and Their Related Developmental Tasks

Infancy and early childhood (up to age 6)	Learning to walk
	Learning to take solid foods
	Learning to talk
	Learning to control the elimination of body wastes
	Learning sex differences and sexual modesty
	Forming concepts and learning language to describe social and physical reality
	Getting ready to read
	Learning to distinguish right and wrong and beginning to develop a conscience
Middle childhood years (6 to 12 years)	Learning physical skills necessary for ordinary games
	Building wholesome attitudes toward oneself as a growing organism
	Learning to get along with age-mates
	Learning appropriate masculine or feminine social role
	Developing fundamental skills in reading, writing, and calculating
	Developing concepts necessary for everyday living
	Developing conscience, morality, and a scale of values
	Achieving personal independence
	Developing attitudes toward social groups and institutions

T A B L E 2-3 (continued)

Adolescent years *(12 to 18 years)*	Achieving new and more mature relations with age-mates of both sexes
	Achieving a masculine or feminine social role
	Accepting one's physique and using the body effectively
	Achieving emotional independence from parents and other adults
	Preparing for marriage and family life
	Preparing for an economic career
	Acquiring a set of values and an ethical system as a guide to behavior—developing an ideology
	Desiring and achieving socially responsible behavior
Early adulthood years *(18 to 30 years)*	Selecting a mate
	Learning to live with a marriage partner
	Starting a family
	Rearing children
	Managing a home
	Getting started in an occupation
	Taking on civic responsibilities
	Finding a congenial social group
Middle years *(30 to 60 years)*	Assisting teenage children to become responsible and happy adults
	Achieving adult civic and social responsibility
	Reaching and maintaining satisfactory performance in one's occupation
	Developing adult leisure-time activities
	Relating to one's spouse as a person
	Accepting and adjusting to the physiological changes of middle age
	Adjusting to aging parents
Later maturity years *(60 years and older)*	Adjusting to decreasing physical strength and health
	Adjusting to retirement and reduced income
	Adjusting to death of one's spouse
	Establishing an explicit affiliation with one's age group
	Adopting and adapting social roles in a flexible way
	Establishing satisfactory physical living arrangements

Social Contact and Interpersonal Relationship Skills

Interaction with other human beings is one of the most basic needs that we have. Our lives revolve around one another. We enjoy the company of other people, and our friends are important in our lives. We think, write, and talk about one another. We have developed both written and unwritten rules that govern the way we eat, dress, travel, and get married. We have organized ourselves into political entities for our own well-being. This process of social interaction strongly influences the standards that we are expected to live by, the attitudes that we have toward others, and the actions that we take in particular circumstances.

Some of the difficulties individuals face in the social sphere of their lives include role conflicts and ineffective relationships with others.

ROLE EXPECTATIONS AND ROLE CONFLICT. We grow up and live in a social/cultural milieu composed of four interrelated groups that strongly influence our attitudes, customs, values, and social mores. These four groups are composed of (1) the members of our immediate home and family environment; (2) our peers and the members of our immediate community (including school, church, and club contacts); (3) the members of our subculture; and (4) the members of the general society that surrounds us (see Figure 2-1). The influence that members of any one group have on us may vary; in general, however, the members of the groups closest to us exert the most direct influence, while the members of the groups furthest away exert the least direct influence.

As members of various groups in this social/cultural milieu, we learn to behave in ways that these groups expect. Thus, we behave in a particular *role* in a way that we are expected to behave. Roles represent the ways we carry out the duties and obligations incumbent in our membership in different groups. Role theory and role expectations are important in understanding the social dynamics of human behavior and the transactions that occur between and among individuals. The major constructs of role theory are that:

- we all occupy one or more positions in our society (for example, mother, aunt, student);
- all of us who occupy a particular position (counselors in a particular setting) behave in distinctive ways that are common to all occupants of that position (all counselors);
- this characteristic way of behaving is expected by other members of society as well as the occupants of that position; and
- the other members of society fall into other distinctive groups that form a network of interrelated groups.

Role expectations are normally quite clear when there is agreement among the members of the occupant group and among the members of other interrelated groups. Role conflicts may occur when there is a lack of agreement

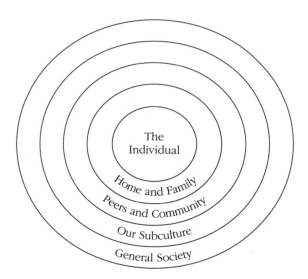

FIGURE 2-1.
Our social/cultural milieu

among group members or among the members of the group and the members of the interrelated groups, or when an individual behaves in a manner that is contrary to the standards expected by the group members.

Effectively functioning individuals may not experience role conflicts, or they may consider these conflicts an inevitable part of life and resolve them in a proactive way by using any of the constructive coping skills discussed previously. Individuals who are functioning less effectively encounter difficulty handling these role conflicts, and they often experience tension, anxiety, and dissatisfaction in the role in which the conflict exists. Furthermore, the less effective person will typically use nonconstructive or reactive adjustment mechanisms in resolving these role conflicts.

LONELINESS, SHYNESS, AND INEFFECTIVE COMMUNICATION SKILLS. In addition to role conflicts, three major difficulties that people may experience in the social sphere of their lives are loneliness, shyness, and an ineffective communication process.

Loneliness has been classified into three distinctive forms by Young (1979). First, there is the brief and momentary feeling of isolation, or a *transient* type of loneliness, that anyone may have. This form of isolation is not a serious problem for most people. Second, there is the *transitional* type of loneliness, which lasts longer; it may be caused by a geographical move, a divorce, or the death of a spouse. This kind of loneliness can cause considerable pain and will last until some coping process is employed. Finally, there is the *chronic* type of loneliness, which is experienced by those who are socially isolated and not meeting their needs for a satisfactory social life. This form of loneliness can be quite serious and is often related to other problems.

Lonely people usually have poor self-regard, a lack of self-confidence, and low self-esteem. Socially isolated students of all ages do poorly academically and experience little satisfaction in school settings (Rice, 1984). Loneliness is also related to a lack of effective social skills, a fear of intimacy, and an irrational thinking process (Weiten, Lloyd, & Lashley, 1991). Socially isolated people want to be liked by others, but they either engage in self-defeating behaviors or withdraw from interacting with others.

Shyness is another major social problem for many people. It can be the cause of social isolation. Children who do not learn how to interact effectively with other children can become lonely and withdrawn (Dworetzky, 1987). Although shyness and loneliness are interrelated, they are not the same. Shy people have difficulty making friends; they are not necessarily lonely. Shy people exercise an extreme amount of caution in social situations, whereas lonely people have a chronic feeling or sense of being alone. According to Zimbardo (1977), shyness tends to be experienced in specific social encounters such as large groups or new situations and with specific groups of people such as strangers and authority figures. Shyness is not considered to be an enduring personality trait. Helping a client deal with shyness is often less difficult than helping someone overcome loneliness.

An important aspect of our interaction with others is effective communication. Good communication occurs when two or more people focus their attention on the same issue at the same time and understand the meaning of the expressed comments. Effective communication is essential for good relationships. Poor communication is a common problem with those who have inadequate and dysfunctional relationships with others. The primary cause of a communication difficulty may be with the speaker, the listener, or both. A major cause of poor communication is lack of careful attention to what is being said. The speaker may communicate self-centered, irrelevant, or tangential information; send different verbal and nonverbal messages; create an interrogative atmosphere; employ distracting nonverbal or paraverbal mannerisms; or not think about the effect the words will have on the listener. The listener may not attend to the speaker or be preoccupied or may selectively listen and hear only parts of the speaker's message. Effective communication requires effective speaking and effective listening. Clients who exhibit poor skills in this area can benefit from an intervention strategy that incorporates communication-skill training.

The effectively functioning person has developed ways to interact with others that are useful, helpful, and supportive. Role conflicts are dealt with in proactive ways, and the effective person manifests positive attitudes, thoughtfulness, and empathic understanding of others. He or she has developed satisfying relationships and effective two-way communication systems with significant other persons in all spheres of life. These communication systems are based on mutual respect, warmth, and understanding. The peson who is functioning less effectively often demonstrates role conflicts, loneliness, shyness, and ineffective communication skills.

During the initial phases of counseling, you should gain some insight into how your client is functioning in human relationships. A client who manifests one or more of the characteristics of the less effective type of person may need an intervention that incorporates three different objectives: obtaining insight into the problem area, improving self-esteem, and/or improving social skills. Treatment programs for clients who have these socially oriented problems frequently include role-playing exercises, assertiveness training, and cognitive-restructuring procedures.

Other Characteristic Attributes

Our ability to function effectively is manifest by other personal attributes and by the characteristic ways that we interact with our environment. The self-actualizing, emotionally healthy, and effectively functioning person manifests positive personal attributes and has a pattern of interaction with the environment that is rich and dynamic. The effective or more fully functioning person exhibits positive attributes and characteristic ways of behaving such as a sense of personal identity and self-worth, a commitment to some meaningful goals, a sense of responsibility for self-control and tolerance, and initiative and spontaneity (Maslow, 1970). Furthermore, the effectively functioning person typically manifests good cognitive skills. These hallmarks of psychological independence and interdependence are amplified in the following sections.

SENSE OF PERSONAL IDENTITY AND SELF-WORTH.   Effectively functioning individuals have a good sense of their own personal identity and positive feelings about themselves. Having this positive self-evaluation enhances their self-concept and enables them to function at high levels. Individuals who function less effectively do not feel good about themselves. Low self-esteem is highly correlated with emotional problems, poor school achievement, and poor interpersonal relationship skills.

COMMITMENT TO MEANINGFUL GOALS.   Self-actualizing persons are aware of societal demands and the necessity to make choices throughout life. Healthy individuals accept this challenge and commit themselves to some meaningful goals or objectives that both enhance the self and contribute to the overall well-being of others. Effective individuals have a good sense of immediate, intermediate, and long-term goals, and they direct their efforts in a meaningful way toward appropriate goals. Less effective individuals are unable to develop plans, or they may be unable to implement the plans they have made.

RESPONSIBLE SELF-CONTROL AND TOLERANCE.   Healthy individuals experience the full range of human desires and emotions and have learned to respond to these feelings in a responsible, self-controlled manner. Individuals who are functioning effectively accept responsibility for their actions and recognize that conflicts often exist between one's personal needs or desires and the needs and desires of others. The effective individual accepts these conflicts as normal and healthy. Individuals who are functioning less effectively often have difficulty with self-control and are frequently intolerant of others.

TAKES INITIATIVE AND REVEALS SPONTANEITY. Self-actualizing individuals tend to be proactive rather than reactive. They anticipate the many problems and concerns that individuals must deal with in the course of life. This anticipatory preparation facilitates problem solving during developmental periods and crisis situations, and fosters positive growth. Effective individuals do not see themselves from static points of view such as being perfect or having all the things they need in life. Rather, they recognize their own imperfections and the transience of acquiring physical things, and that all humans are in a dynamic state of growing and becoming. Their actions tend to be reasonably spontaneous and in concert with highly prized values (Fromm, 1987; Rogers, 1972). Individuals who are functioning less effectively tend to be reactive rather than proactive. This reactive process is a defensive and protective way of behaving. Ineffective individuals frequently lack spontaneity and initiative.

EFFECTIVE COGNITIVE SKILLS. Functioning effectively in the cognitive domain is usually manifest in good learning patterns, accurate perceptions of reality, sound logical and inferential thinking processes, good decision-making skills, and intellectual openness.

> *Good learning patterns.* Learning is a complex process that strongly influences the behavior of individuals. Learning theorists state that we learn through observational learning, classical and operant conditioning, and through a cognitive-information-process model that includes sensory stimulation, examining available information, selecting and weighting the importance of the data, and forming meaningful associations (Hilgard & Bower, 1975; Kintsch, 1982; Wickelgren, 1974b). Effective individuals have mastered good learning habits; they realize the importance of multisensory input, repetition, and selectivity when exposed to information and are able to categorize this information in an organized way.
> *Accurate perceptions of reality.* An important element in our cognitive functioning is how we select, organize, and interpret the information we have obtained. Effective people can select appropriate information from an array of data, organize the various pieces of this information into a meaningful hierarchical order, and interpret the information in ways that facilitate their awareness of reality (Dember & Warm, 1979; Hochberg, 1978).
> *Sound logical and inferential thinking processes.* An important cognitive skill is our ability to think logically and reach sound conclusions on the basis of given information. Effective individuals can think both

deductively and inductively. They can reach meaningful conclusions or make reasonable inferences about new and previously unknown objects and events on the basis of previous observations of similar phenomena (Kintsch, 1982; Wickelgren, 1974a).

Good decision-making skills. The ability to make a decision has several cognitive components. The process involves

(1) becoming aware of the decision-making process,
(2) identifying the problem,
(3) determining appropriate goals,
(4) describing all the important factors related to the problem,
(5) identifying alternatives and the consequences of these alternatives,
(6) judging the desirability of the alternatives, and
(7) selecting, implementing, and evaluating the chosen alternative.

Effective individuals have manifested a history of decision making that has enhanced their quality of life in one or more of the major spheres of life (Krumboltz & Thoresen, 1976; Wickelgren, 1974b).

Intellectual openness. Intellectual openness is the ability to think about things in a fresh or new way, to conceptualize new behaviors, and to try them out intellectually. The effective individual recognizes that many aspects of life must be thought of in terms of probabilities and chance factors rather than absolutes and is willing to take some intellectual risks (Coan, 1974; Fromm, 1987; Rogers, 1972).

Counselors with clients who manifest poor images, who lack meaningful goals, who have limited self-control, or whose cognitive functioning is ineffective, must develop treatment programs that focus on improving these aspects of their clients' personality. A variety of techniques for accomplishing these goals is available. Chapters 4, 5, and 6 outline several important cognitive, affective, and behavioral strategies that can be used in working with clients who have these concerns.

SPECIAL PROBLEMS THAT IMPEDE EFFECTIVE FUNCTIONING

Several distinctive problems may impede the effective functioning of individuals. Four of these problems are handicapping conditions, serious vocational problems, substance-abuse problems, and psychological disorders.

Handicapping Conditions

The adjustment and behavioral patterns of individuals are influenced by personal characteristics such as physical limitations caused by birth defects, sickness, or accidents; intellectual impairments brought about by neurological damage or learning disorders; and other concerns related to health problems. Effectively functioning individuals will react to these blockages in proactive ways,

whereas persons functioning at less effective levels will be more reactive in their responses.

DISABILITIES. A person who is disabled has a permanent condition caused by an accident, disease, or congenital problem that interferes with that person's ability to function. Counseling clients with disabilities frequently involves helping them make realistic assessments of their strengths and limitations and assisting them in learning how to adjust and compensate for any known limitation. Because disabling conditions are blockages or frustrations that cause emotional feelings of anger, the counseling process should include helping the client assess his or her coping skills. When necessary, the client should be given the appropriate assistance to overcome nonconstructive adjustment mechanisms and develop more constructive coping strategies.

LEARNING DIFFICULTIES. Individuals of average or above average intelligence who encounter serious difficulties in learning are considered learning disabled. They often see themselves as slow or incompetent. Although the origin of any learning disability is assumed to be neurological, it is often impossible to determine the specific cause of a learning problem for a particular client. Consequently, when working with individuals who have learning disorders, it is appropriate to focus not on the cause of the problem but rather to pinpoint specific behavioral learning deficits and formulate appropriate remedial programs. Typically, clients with learning disabilities have perceptual learning deficits that have concomitant effects on their educational progress, socialization, and self-esteem.

Counselors who work with individuals who have some physical or neurological damage and either a physical or learning disability must work cooperatively with a team of other knowledgeable persons. This team may include diagnostic experts, classroom teachers, special-education teachers, specialists in reading or speech and hearing, rehabilitation counselors, and the parents of the client. The counseling process typically focuses on helping clients improve their self-images, change their illogical thinking processes, improve their study skills and habits, and enhance their social skills.

Serious Vocational Concerns

The occupation a person chooses has a profound influence on all aspects of his or her life. It may determine who you marry, how much money you earn, where you live, and how you spend your leisure time. The notion that an occupational choice is made at a particular point in life is misleading. Occupational selection is not really a onetime event; it is a lifelong process involving developing abilities, interests, and values, and making numerous and varied educational and vocational decisions.

Vocational counseling experts such as Tolbert (1980) and Zunker (1990) maintain that decisions about occupational and career choice require individuals to have a sound understanding of themselves, a comprehensive knowledge about the occupational opportunities available, and a thoughtful career-planning process. There is clear evidence that not all individuals function effectively in the vocational aspects of their lives. Their behavior may be considered vocationally maladaptive. Although no universal taxonomy of vocationally maladaptive behavior has been developed, those who are having vocational problems may belong to one or more of the following groups.

THOSE WHO ARE UNABLE TO MAKE A CAREER DECISION.   This group consists of individuals who are at the appropriate age to make a vocational decision but who cannot make one. There are several different undecided groups (Osipow, 1980): (1) those who have low self-confidence or high anxiety; (2) those who believe that there are serious barriers to their appropriate vocational choice; (3) those who have difficulty deciding among alternatives; and (4) those who lack sufficient information about themselves or the occupational world to make a sound decision.

THOSE WHO HAVE DEFECTIVE WORK PERSONALITIES.   Neff (1985) has suggested that there are five distinctive patterns among defective work personalities. These individuals tend to have problems with independence, acceptance of the work ethic, interpersonal relationships, authority figures, and emotional stability. The five types are:

1. The *dependent*—individuals whose major response to the work demand is dependency and childlike reliance on others for support.
2. The *impulsive*—people whose major response to the work demand is indifference, who lack a concept of responsibility, and whose impulse gratification is strong.
3. The *socially naive*—persons whose response to the work demand is marked by naiveté and who need experience to gain skills in the social mores and customs of work.
4. The *hostile*—individuals whose response to the work demand is hostility and aggression and who can work alone well but whose anger is easily aroused.
5. The *fearful*—those whose major response to the demand to work is fear and anxiety; they have a history of learned helplessness.

According to Neff, potential workers who manifest one of these five patterns need a carefully developed rehabilitation program to ensure their employability.

THOSE WHO ARE STRUCTURALLY UNEMPLOYED.   The structurally unemployed are individuals who are not employed because their skills do not match the requirements for the available jobs. They include individuals who are inadequately prepared or inappropriately educated for the demands of the available jobs and those who are geographically immobile, inexperienced, or handicapped (Herr & Cramer, 1988). Structurally unemployed people are often very discouraged and may not be considered part of the labor force because they do not actively seek work.

THOSE WHO ARE UNDEREMPLOYED.   The underemployed are individuals who are either working part-time but seeking full-time employment or working in positions that require much less education and training than they have (Weiten, Lloyd, & Lashley, 1991). Underemployed workers are probably also dissatisfied because of their unmet needs.

THOSE WHO ARE VOCATIONALLY DISSATISFIED.   This group represents a wide spectrum of workers whose present jobs fail to provide important economic, psychological, or sociological satisfactions. Nevertheless, many dissatisfied workers remain in their positions for one reason or another. According to Herr and Cramer (1988) some workers who are extremely dissatisfied with their jobs may suffer from *burnout,* which is normally caused by stress factors related to the job.

Clients who are vocationally maladaptive can benefit from vocational counseling. The vocational-counseling process will typically help clients develop more insight, acquire some necessary vocational skills, receive emotional support during any retraining process, learn the techniques necessary to explore various occupational opportunities, learn how to search out potential jobs, and embark on appropriate job-seeking endeavors. The specific treatment strategies used will vary depending on the unique needs of a particular client.

Substance-Abuse Problems

Substance abuse involves the use of alcohol or another drug to such an extent that one's personal and social functioning are impaired. All drugs have a physiological and a psychological effect. They induce a chemical reaction in the cells of the brain that affects the nervous system and causes changes in perceptions, emotions, and behavioral patterns. The major difficulties in using any drug are the potential for physical and psychological dependence, the potential for causing serious harm to oneself and others, and the enormous costs to society.

Although alcohol is similar to other drugs in many respects, the issues associated with each of the following substances are different enough to be considered as separate concerns.

ALCOHOL ABUSE.   Alcohol is widely used throughout the world. It is not considered an illegal substance, nor does it require a prescription as do many other drugs. Although there are a variety of methods for classifying people who drink, a very useful way is to group them by their degree of alcoholic dependency (Forrest, 1984; Peer, Lindsey, & Newman, 1982). *Social drinkers* are not dependent upon alcohol. They consume a moderate amount of alcohol, but they do not feel a compulsion to drink and they can accept or refuse a drink. *Problem drinkers* have a psychological dependence on alcohol and consume a considerable amount, often daily, to relieve tension. They usually reserve their drinking, however, for times that do not interfere with their work or their other major responsibilities. *Alcoholics* have both a psychological and a physiological dependence on alcohol. They usually consume a large amount daily, and they need a drink to reduce the stress they feel and to cope with the situations they encounter each day. Other authorities have classified alcoholics into further subdivisions (Jellinek, 1960).

Addiction to alcohol is one of the most serious health and social problems of our era (Gitlow & Peyser, 1980; Landy, 1987; Quayle, 1985). This maladaptive behavior is believed to be significantly related to one-eighth of our national health costs; 50 percent of all automobile fatalities; 60 percent of all homicides; and high accident and absentee rates and low productivity in industry. Furthermore, a large number of diseases (cirrhosis of the lever, pancreas disorders, and various nutritional problems) are specifically related to alcohol addiction. It is considered the third leading cause of death in the United States, after heart disease and cancer.

Alcoholics often deny that they have a drinking problem, and they frequently resist suggestions that they go for treatment. However, this maladaptive behavior pattern can be altered. Most treatment programs involve individual counseling, family counseling, and involvement in a support group such as Alcoholics Anonymous, Al-Anon, or Alateen. In cases where the alcoholism is severe, a client may be hospitalized to overcome withdrawal symptoms.

The individual counseling sessions focus on helping clients abstain from drinking, improve their feelings of self-worth, and learn to cope successfully with the stressful conditions in their lives. The family counseling sessions are designed to help families understand and deal with their feelings of denial and shame, improve their dysfunctional communication patterns, and modify behaviors that have unwittingly reinforced the drinking process. Many treatment programs strongly encourage clients to join Alcoholics Anonymous, their spouses to meet with an Al-Anon group, and their teenage children to become associated with an Alateen group. These groups not only provide emotional support to the members, they also help the individual and the family deal with the temptations of alcoholism and the problems of remaining sober.

DRUG ABUSE. The use and abuse of psychoactive substances has increased significantly in the last several years and has become a problem in all parts of the nation (Ray, 1982). Addicts may be found in all age groups and in all classes of society; however, drug abuse appears to be more serious among adolescents and young adults. Furthermore, a considerable amount of illegal behavior is involved in the abuse of drugs. Elaborate smuggling and illicit distribution are required to get the drugs to their users, and serious addicts frequently turn to crime in order to obtain the money to support their habits. Drugs are classified by their effect on the user, by chemical composition, or by the source of the drug and can be categorized in the following way:

Narcotics (analgesics). A narcotic is a drug that has both a pain-relieving (analgesic) and a sleep-inducing (sedative) effect. Because these drugs are made from the opium or poppy plant, which is grown in hot, dry climates, they are also referred to as *opiates*. Well-known narcotics include codeine, heroin, and morphine. Narcotics can also be manufactured. Examples of synthesized narcotics are methadone, meperidine (Demerol), and oxycodone (Percodan). Narcotics act on the nervous system, producing drowsiness, a feeling of euphoria, and a sense of detachment from pain. Using this type of drug often causes physical and psychological dependence, a lowering of social motivation, and the possibility of infections from nonsterile injections.

Hallucinogens (psychedelics). Hallucinogens are drugs that do not have any known medical value. However, they do have an extremely powerful effect on emotional, intellectual, and motor functioning. Representative examples are mescaline (obtained from the peyote cactus), psilocybin (derived from a rare mushroom plant), dimethyltriptamine (DMT), and lysergic acid diethylamide (LSD). This type of drug acts on the central nervous system to alter auditory, visual, and tactile perceptions. It tends to induce a highly euphoric or extremely depressed emotional state and seriously impairs memory and problem-solving skills. Potential hazards from the use of this sort of drug are changes in mood; feelings of intense anxiety, panic, or depression; and the inability to distinguish between reality and fantasy.

Stimulants (speed). A stimulant is a drug that elevates the mood, emotions, alertness, and physical activity of the user. Sometimes referred to as "speed" or "uppers," the primary stimulants are amphetamines (Dexedrine and Benzedrine), cocaine, and caffeine. These drugs stimulate the central nervous system, thereby increasing alertness, suppressing appetite temporarily, and increasing optimism. The major dangers in using stimulants are psychological dependence and a deterioration of physical health. Heavy use can cause aggressiveness, confusion, and psychotic-like behavior.

Sedatives. A sedative is a drug that has both a sleep-inducing and a depressant effect. The opposite of the stimulants, sedatives tend to relax and soothe, and combat anxiety and restlessness. Sometimes referred to as "downers," examples of sedatives include the barbiturates (Amytal, Luminal, Nembutal, Seconal), the chlorpromazines (Thorazine and Serpasil), the meprobamates (Miltown and Equanil), chlordiazepoxide (Librium), and diazepam (Valium). Large doses have similar effects to those of consuming large amounts of alcohol (euphoria, slurred speech, and poor motor coordination). The major risks in using sedatives are physical and psychological dependence and the strong possibility of physical injury due to poor motor coordination.

Cannabis sativa (marijuana). Marijuana, hashish, and tetrahydrocannabinols (THC) are obtained from the *Cannabis sativa* or hemp plant, which can be grown easily in many parts of the world. Marijuana is a mixture of the flowers, leaves, and stems of the plant; hashish is extracted from the resins; and THC is the active chemical in the plant. The major physiological effects of the drug are an increase in heart rate and a reddening of the eyes. Psychologically the drug tends to produce a mild state of intoxication, induce a sense of relaxation and euphoria, distort the sense of taste and touch, and impair memory and reaction time.

Substance-abuse counseling is a distinct specialization. Treatment for drug abusers typically involves placement in a highly structured residential facility, which is often staffed by ex-addicts. These programs typically involve medical treatment for the withdrawal process and both individual and group counseling sessions based on the principles of behavioral counseling. Although many counselors have little direct experience with the rehabilitation of substance abusers, this problem is so widespread in our society that all counselors need to be knowledgeable about substance abuse for at least three reasons. First, counselors have an important role in primary prevention (Baker & Shaw, 1987); second, counselors must know how to make referrals to appropriate substance-abuse specialists; and third, counselors working in schools and other institutional settings often need to provide supportive counseling for clients receiving specialized counseling outside the institutional setting.

Psychological Disorders

Individuals who exhibit extreme forms of personal distress, unrealistic or irrational behavior, social deviance, or other manifestations of maladaptive behavior are considered to be mentally ill, abnormal, psychologically deviant, or pathological. Those who suffer from a psychological disorder have been classified into various diagnostic categories. The official reference for this grouping is the revised third edition of the *Diagnostic and Statistical Manual of Mental Disorders* (*DSM-III-R*) published in 1987.

DISORDERS USUALLY FIRST EVIDENT IN INFANCY, CHILDHOOD, OR ADOLESCENCE.   This diagnostic category includes several major disorders that appear in infancy, childhood, or adolescence. People who have a *developmental* disorder have experienced difficulties in acquiring cognitive, language, motor, or social skills. Those who exhibit a *disruptive-behavior* disorder manifest serious socially disruptive behavior. Those who reveal an *anxiety* disorder in childhood and adolescence have an extremely high level of apprehension. And those who have an *eating* disorder demonstrate gross disturbances in eating patterns. Other groupings in this diagnostic category include disorders manifested by tics, disturbances in gender identity, elimination dysfunctions, and speech disorders not elsewhere classified.

ORGANIC MENTAL SYNDROMES AND DISORDERS.   Individuals with these disorders manifest either a temporary or a permanent impairment of the brain. The most prevalent organic mental syndromes are delirium, dementia, intoxication, and withdrawal. Individuals who have a *delirium* disorder experience difficulty sustaining or shifting attention and maintaining a coherent thought pattern. Persons with a *dementia* disorder exhibit difficulty with memory, abstract thinking, and making sound judgments. Those with an *intoxication* disorder reveal ill-advised behavior patterns, take a psychoactive substance regularly, and have an impaired central nervous system. Individuals who have a *withdrawal* disorder have stopped or substantially reduced their intake of a psychoactive substance.

PSYCHOACTIVE-SUBSTANCE-USE DISORDERS.   Substance-use disorders are characterized by individuals who use a substance frequently, suffer some noticeable impairment because of its use, become dependent upon the substance, and manifest an impaired ability to control their use of a particular substance.

SCHIZOPHRENIC DISORDERS.   Schizophrenic disorders are manifested by individuals who have severe thought disturbances and who appear to be out of contact with reality. There are several major schizophrenic disorders: Persons who have a *paranoid-type* disorder experience delusions of persecution. Those who manifest a *catatonic-type* disorder have rather extreme motor disturbances. Individuals who have a *disorganized-type* disorder behave in a rather silly and disorganized way. Persons who have an *undifferentiated-type* disorder display schizophrenic symptoms but cannot be classified into one of the previously mentioned categories. Finally, those who have a *residual-type* disorder have had at least one schizophrenic episode but presently do not have any obvious psychotic symptoms.

DELUSIONAL (PARANOID) DISORDER.   This disorder is exhibited by individuals who have delusions of persecution or grandeur but who do not have the severe deterioration seen in paranoid schizrenia.

PSYCHOTIC DISORDERS NOT ELSEWHERE CLASSIFIED.   Disorders fall into this diagnostic category when they cannot be classified into one of the other psychotic categories. There are four major groups in this category: Persons who manifest a *brief-reactive-psychosis* disorder have had brief episodes of psychotic symptoms that may last anywhere from a few hours to no longer than one month. Individuals who have a *schizophreniform* disorder display schizophrenic symptoms that last less than six months. Those who have a *schizoaffective* disorder exhibit symptoms of schizophrenia or mood disorders but do not meet the criteria for either classification. Persons who have an *induced-psychotic* disorder experience delusions that have developed from being in a close relationship with another person who has a psychotic disorder.

MOOD DISORDERS.   Mood disorders are marked by persistent disturbances in mood and emotional tone. There are two major categories of mood disorders. First, persons who have a *depressive* disorder exhibit feelings of sadness, dejection, and despair that have become paramount in their lives. In severe cases of depression, delusions and hallucinations may occur. Second, individuals who have a *bipolar mood* disorder experience depression and its bipolar opposite, a euphoric or manic state.

ANXIETY DISORDERS.   Anxiety disorders are psychological disturbances in which an individual's behavior is dominated by feelings of tension, fear, apprehension, and avoidance. There are several major anxiety disorders: Persons who have a *panic* disorder manifest recurrent panic attacks. Individuals with a *phobic* disorder experience persistent and irrational fears of objects or situations that present no real danger. Those who have an *obsessive-compulsive* disorder have unwanted thoughts, ideas, or impulses to engage in certain repetitive actions that appear absurd and useless. Persons who exhibit a *posttraumatic stress* disorder have had a very stressful experience and maintain a strong reaction to this event for relatively long periods. Finally, individuals who manifest a *generalized anxiety* disorder have a chronic and high level of apprehension not related to any known threat.

SOMATOFORM DISORDERS.   Somatoform disorders are manifested by physical symptoms for which no authentic organic basis can be discovered. There are several major somatoform disorders. Persons who have a *body-dysmorphic* disorder are preoccupied with some imagined defect in their physical appearance. Individuals who exhibit a *conversion* disorder experience a significant loss of some physiological function that appears to be an expression of a psychological need. Those who have *hypochondriasis* are excessively worried about the possibility of contracting a serious illness. People who manifest a *somatization* disorder have a long and recurring history of chronic but diverse medical complaints. Individuals who exhibit a *somatization pain* disorder are preoccupied with a pain of unknown origin. Finally, those with an *undifferentiated somatoform* disorder have symptoms similar to the somatization disorder but do not meet the requirements for this classification.

DISSOCIATIVE DISORDERS.   Dissociative disorders are those disturbances manifested by individuals who lose contact with portions of their consciousness or memory. There are three major dissociative disorders. First, persons who have a *multiple-personality* disorder appear to have two or more distinct personalities. Second, individuals who exhibit a *psychogenic fugue* disorder forget who they are, relocate geographically, change occupations, and develop new identities. Third, those who exhibit a *psychogenic amnesia* disorder have a sudden and extensive memory loss. Finally, individuals who experience a *depersonalization* disorder temporarily lose or modify their own sense of personal identity.

SEXUAL DISORDERS.   There are two major classifications of sexual disorders. Persons who manifest a *paraphilia* disorder have recurring sexual urges, fantasies, and/or behaviors that are not part of the normal sexual arousal process. Individuals who have a *sexual dysfunctional* disorder experience difficulties with appropriate responses during sexual relations.

SLEEP DISORDERS.   There are two major subgroups of sleep disorders: dyssomnia and parasomnia. Those who suffer from *dyssomnia* experience difficulty in the amount of time that they sleep or the appropriateness of the time when they fall asleep, or in not feeling rested after an appropriate period of sleep. Individuals who suffer from *parasomnia* have abnormal occurrences during their sleep such as frightening dreams or sleepwalking.

FACTITIOUS DISORDERS. Individuals who manifest disorders in this classification have physical and/or psychological symptoms of illness that are deliberately produced to satisfy some need. Although these symptoms are purposeful, they are not under the conscious control of the individual.

IMPULSE-CONTROL DISORDERS. These disorders are exhibited by those who are unable to control an impulse, drive, or temptation to act in some way harmful to themselves or to another person. There are five major types of disorders in this category. First, individuals who have *intermittent explosive* disorder may either aggressively assault other individuals or destroy property. Second, persons with a *kleptomania* disorder cannot resist the tendency to steal property that they do not need. Third, those with a *pathological gambling* disorder cannot resist the temptation to gamble. Fourth, individuals who have a *pyromania* disorder cannot resist the impulse to set fires. Finally, those who have a *trichotillomania* disorder cannot overcome their temptation to pull out their own hair.

ADJUSTMENT DISORDERS. Individuals who have an *adjustment* disorder experience maladaptive reactions to some stressful condition. The adjustment difficulty may become evident in one or more areas of life such as in school, at work, or in one's social life.

PERSONALITY DISORDERS. When an individual's personality traits become inflexible and cause functional impairment or subjective distress, they become personality disorders. These disorders often become manifest during the teenage years and may continue throughout one's adult life. The *DSM-III-R* lists 11 personality disorders grouped into three clusters. The first group, the *odd or eccentric* disorders, contains the paranoid, the schizoid, and the schizotypal personality disorders. The second group, the *dramatic, emotional, or erratic* disorders, includes the antisocial, the borderline, the histrionic, and the narcissistic personality disorders The third group, the *anxious or fearful* disorders, comprises the avoidant, the dependent, the obsessive-compulsive, and the passive-aggressive personality disorders.

‖ SUMMARY

‖ This chapter has briefly outlined several ways that effective individuals differ from ineffective ones. After studying this chapter you should be able to describe the major needs individuals have, discuss the role of stress and the emotions in the coping process, outline the major developmental concerns, indicate some important social interaction and relationship skills, and name some other major characteristics of effectively functioning individuals. In addition, you should have gained some understanding of individuals with special concerns, such as those with handicapping conditions, serious vocational concerns, substance-abuse problems, and psychological disorders. To attain a high level of understanding of these concerns, in-depth study of these topics is required.

REFERENCES

Adler, A. (1964). *Social interest: A challenge to mankind.* New York: Capricorn.

American Psychiatric Association. (1987). *Diagnostic and statistical manual of mental disorders* (3rd ed., rev.). Washington, DC: Author.

Baker, S. B., & Shaw, M. C. (1987). *Improving counseling through primary prevention.* Columbus, OH: Merrill.

Bernard, H. W., & Huckins, W. C. (1976). *Dynamics of personal adjustment* (2nd ed.). Boston: Holbrook.

Biehler, R. F. (1976). *Childhood development: An introduction.* Boston: Houghton Mifflin.

Coan, R. W. (1974). *The optimal personality.* New York: Columbia University Press.

Coleman, J. C. (1979). *Contemporary psychology and effective behavior.* Glenview, IL: Scott, Foresman.

Conger, J. J. (1983). *Adolescence and youth: Psychological development in a changing world* (3rd ed.). New York: Harper & Row.

Dember, W. N., & Warm, J. S. (1979). *Psychology of perception* (2nd ed.). New York: Holt, Rinehart & Winston.

Dinkmeyer, D. C., Sr., Dinkmeyer, D. C., Jr., & Sperry, L. (1987). *Adlerian counseling and psychotherapy* (2nd ed.). Columbus, OH: Merrill.

Dreikurs, R. R., & Soltz, V. (1964). *Children: The challenge.* New York: Hawthorne.

Dworetzky, J. P. (1987). *Introduction to child development* (3rd ed.). St. Paul, MN: West Publishing.

Elson, S. E. (1979). Recent approaches to counseling: Gestalt, transactional analysis, and personality theories. In H. M. Burks & B. Stefflre (Eds.), *Theories of counseling* (3rd ed.) (pp. 254–316). New York: McGraw-Hill.

Erikson, E. H. (1963). *Childhood and society* (2nd ed.). New York: Norton.

Erikson, E. H. (1968). *Youth, identity, and crisis.* New York: Norton.

Forrest, G. G. (1984). *Intensive psychotherapy of alcoholism.* Springfield, IL: Charles C Thomas.

Freud, A. (1966). *The ego and the mechanisms of defense* (rev. ed.). New York: International Universities Press.

Fromm, E. (1987). *To have or to be.* New York: Bantam Books.

Gitlow, S. E., & Peyser, H. S. (Eds.). (1980). *Alcoholism: A practical treatment guide.* New York: Grune & Stratton.

Goodstein, L. D., & Lanyon, R. I. (1979). *Adjustment, behavior, and personality* (2nd ed.). Reading, MA: Addison-Wesley.

Havighurst, R. J. (1972). *Developmental tasks and education* (3rd ed.). New York: David McKay.

Herr, E. L., & Cramer, S. H. (1988). *Career guidance and counseling through the life span: Systematic approaches* (3rd ed.). Boston: Scott, Foresman.

Hershenson, D. B., & Power, P. W. (1987). *Mental health counseling: Theory and practice.* Elmsford, NY: Pergamon Press.

Hilgard, E. R., & Bower, G. H. (1975). *Theories of learning* (4th ed.). New York: Appleton-Century-Crofts.

Hochberg, J. (1978). *Perception* (2nd ed.). Englewood Cliffs, NJ: Prentice-Hall.

Jellinek, E. M. (1960). *The disease of alcoholism.* New Haven, CT: Hillhouse Press.

Kagan, J., & Havemann, E. (1988). *Psychology: An introduction* (6th ed.). New York: Harcourt Brace Jovanovich.

Kintsch, W. (1982). *Memory and cognition.* New York: Krieger.

Kohlberg, L., & Kramer, R. (1969). Continuities and discontinuities in child and adult moral development. *Human Development, 12,* 93–120.

Krumboltz, J. D., & Thoresen, C. E. (Eds.). (1976). *Counseling methods.* New York: Holt, Rinehart & Winston.

Landy, F. J. (1987). *Psychology: The science of people* (2nd ed.). Englewood Cliffs, NJ: Prentice-Hall.

Lerner, R. M. (1986). *Concepts and theories of human development* (2nd ed.). New York: McGraw-Hill.

London, H., & Exner, J. E. (1975). *Dimensions of personality.* New York: Wiley.

Maddi, S. R. (1989). *Personality theories: A comparative analysis* (5th ed.). Pacific Grove, CA: Brooks/Cole.

Maslow, A. H. (1970). *Motivation and personality* (2nd ed.). New York: Harper & Row.

Moos, R. H., & Billings, A. G. (1982). Conceptualizing and measuring coping resources and processes. In L. Goldberger & S. Breznitz (Eds.), *Handbook of stress: Theoretical and clinical aspects* (pp. 212–230). New York: Free Press.

Neff, W. (1985). *Work and human behavior* (3rd ed.). New York: Aldine-Atherton.

Nikelly, A. G. (1977). *Achieving competence and fulfillment.* Pacific Grove, CA: Brooks/Cole.

Osipow, S. (1980). *Manual for the career decision scale.* Columbus, OH: Marathon Consulting Press.

Peer, G. G., Lindsey, A. K., & Newman, P. A. (1982). Alcoholism as a stage phenomena: A frame of reference for counselors. *Personnel and Guidance Journal, 60,* 465–469.

Piaget, J. (1966). *The origins of intelligence in children.* New York: International Universities Press.

Quayle, D. (1985). American productivity: The devastating effect of alcoholism and drug abuse. In J. F. Dickman, W. G. Emener, Jr., & W. S. Hutchison, Jr. (Eds.), *Counseling the troubled person in industry* (pp. 20–29). Springfield, IL: Charles C Thomas.

Ray, O. S. (1982). *Drugs, society, and human behavior* (3rd ed.). St. Louis: C. V. Mosby.

Rice, P. F. (1984). *The adolescent: Development, relationships, and culture* (4th ed.). Boston: Allyn & Bacon.

Rogers, C. R. (1959). A theory of therapy, personality, and interpersonal relationships as developed in the client-centered framework. In S. Koch (Ed.), *Psychology: A study of science* (Vol. 3, pp. 184–256). New York: McGraw-Hill.

Rogers, C. R. (1972). *On becoming a person.* Boston: Houghton Mifflin.

Skinner, B. F. (1953). *Science and human behavior.* New York: Macmillan.

Sweeny, T. J. (1989). *Adlerian counseling: A practical approach for a new decade* (3rd ed.). Muncie, IN: Accelerated Development.

Tolbert, E. L. (1980). *Counseling for career development* (2nd ed.). Boston: Houghton Mifflin.

Torrance, E. P. (1965). *Constructive behavior: Stress, personality, and mental health.* Belmont, CA: Wadsworth.

Weiner, B. (Ed.). (1974). *Cognitive views of human motivation.* New York: Academic Press.

Weiten, W., Lloyd, M. A., & Lashley, R. L. (1991). *Psychology applied to modern life: Adjustment in the 90s* (3rd ed.). Pacific Grove, CA: Brooks/Cole.

Wickelgren, W. A. (1974a). *How to solve problems.* San Francisco: W. H. Freeman.

Wickelgren, W. A. (1974b). *Learning and memory.* Englewood Cliffs, NJ: Prentice-Hall.

Wollams, S., & Brown, M. (1979). *The total handbook of transactional analysis.* Englewood Cliffs, NJ: Prentice-Hall.

Young, J. (1979, September). *Cognitive therapy and loneliness.* Paper presented at the meeting of the American Psychological Association, New York, NY.

Zimbardo, P. G. (1977). *Shyness: What it is, what to do about it.* Reading, MA: Addison-Wesley.

Zunker, V. G. (1990). *Career counseling: Applied concepts of life planning* (3rd ed.). Pacific Grove, CA: Brooks/Cole.

3 || *Transitional Stages in the Counseling Process*

INTRODUCTION

This chapter outlines five important counseling stages and describes the distinctive activities that occur in each of these stages. This is followed by a discussion of the characteristics of reluctant and resistant clients, outlining some of the ways to deal with these clients. Finally, important sources of information about clients are reviewed. After studying this chapter, you should be able to:

- identify the five transitional counseling stages;
- discuss the reasons for calling counseling stages transitional, overlapping, and continuous;
- describe the major characteristics or themes found in each counseling stage;
- outline some of the ways you should deal with reluctant and resistant clients; and
- name five different sources that counselors may use to obtain information about clients and describe the advantages and limitations of each one.

THE TRANSITIONAL STAGES

Counseling is a developmental process; it has a beginning, progresses through an orderly, transitional sequence, and has an ending. This developmental process can be described in terms of the following five stages:

- relationship-building stage—developing the foundation for a sound collaborative working association;
- exploratory stage—examining and understanding the client and his or her frame of reference;

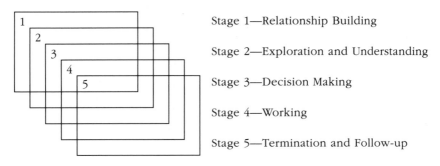

Stage 1—Relationship Building

Stage 2—Exploration and Understanding

Stage 3—Decision Making

Stage 4—Working

Stage 5—Termination and Follow-up

FIGURE 3-1.
The five transitional counseling stages

- decision-making stage—formulating a counseling goal and an intervention strategy;
- working stage—expending effort to ameliorate the situation or solve the problem; and
- termination stage—concluding the counseling process.

The characteristics of the counseling process have been discussed by a number of authors (Baruth & Huber, 1985; Brammer, 1988; Egan, 1990; Hansen, Stevic, & Warner, 1986; Ivey, Ivey, & Simek-Downing, 1987; Okun, 1987; Patterson & Eisenberg, 1983). An analysis of their writings indicates that even though there is no agreement on the number of phases or stages, the names of these stages, nor the characteristics of each stage, there is a consensus that the counseling process moves through a sequential, orderly progression of stages.

It is the viewpoint of this writer that the five stages of counseling are developmental in nature and therefore transitional, overlapping, and continuous rather than rigid, discrete, and discontinuous (see Figure 3-1). This transitional concept implies that, while each of these stages has its own theme and developmental tasks, elements of any stage may be evident at any other stage. The particular counseling stage that one is in can be discerned by the modal or dominant activities of that time span. Furthermore, the emphasis or major theme of each stage must be accomplished to a minimal degree before counseling can progress to the next stage. If some of the developmental tasks of the prior stages have not been accomplished, or if they were minimally met, then the counseling process will be somewhat constrained and limited until such time as these tasks are adequately performed. The counseling process may be recycled at any given point in counseling—that is, at any point the counselor may go back and focus on the developmental tasks of an earlier stage to enrich the process or overcome any difficulties or blockages. This concept of stages is similar to the structured stage concepts outlined by developmental-stage theories (Lerner, 1986; Turiel, 1969). When a particular concern of a client has been resolved, the process may be repeated to focus on another unresolved issue faced by the client.

Counseling, by its very nature, often does not progress on a steady, even, time-predictable course. The character, duration, and intensity of the counselor/ client interaction in different counseling cases varies considerably. Consequently, the character, duration, and intensity of each of these stages also varies. Occasionally, all five stages may occur in one counseling session; however, they normally take a minimum of several sessions and often take many, many sessions. Some counseling cases require considerable time on one or two stages and little time on other stages. Other counseling cases will require considerable time on all stages.

The tasks outlined in each of these five stages appear to be pantheoretical or orientation free; therefore all types of counseling, regardless of the approach, the problem area, or the intervention used, should progress through these stages. The specific counseling tasks associated with each stage are outlined in the following sections. As counselors-in-training, you should become aware of these tasks; realize that the burden of accomplishing these tasks is on you, the counselor, the trained professional, and not on the client, the person in need; and try to develop your skills in accomplishing these tasks in a logical sequence.

The Relationship-Building Stage

When the counseling process begins, the counselor and the client ordinarily do not know one another, so the foundation for a collaborative working relationship must be developed. Building this relationship is the focus of the initial counseling stage, and it is achieved through four interrelated tasks: (1) establishing the facilitative conditions; (2) determining the initial counseling goals; (3) structuring the relationship; and (4) exemplifying ethical standards. This initial stage is normally relatively short, ranging from a few minutes to several sessions.

ESTABLISHING FACILITATIVE CONDITIONS.   At the beginning of this process clients are frequently unsure of themselves, unsure about you, and most of all, unsure about the counseling process. Furthermore, clients may even be unaware of their major concerns. This cautiousness means that clients may be quite guarded in what they say. Clients need to feel a sense of trust, genuineness, and respect. As the counselor you must take the responsibility for providing the conditions that will facilitate open, honest, and complete communication. Clients need to feel a sense of being at ease in the relationship and of being accepted by you, and to feel that they are understood in appropriate ways. You need to communicate an interest and positive regard for clients and an understanding and respect for their feelings, attitudes, and concerns.

These attributes are behaviorally communicated by:

- being fully present to the client and focusing on the client as a person of real value and worth;
- listening intensively and attempting to identify the client's underlying feelings and thoughts; and
- responding to the client with high-level attending responses.

These high-level responses communicate to the client that he or she has been understood; they help the client feel more comfortable in the relationship; and they encourage the client to express more about his or her concerns.

DETERMINING INITIAL COUNSELING GOALS.   Clients come to counselors for a variety of reasons: to resolve a conflict, solve a problem, or obtain some help in becoming a more effectively functioning person. They may be extremely aware of their problems, or rather vague and uncertain about why they feel the way they do or even why they came to the counselor. A client may present an initial problem that is not his or her major concern. Motivation to work on the solution to the problem may be weak, moderate, or strong. In any case, some mutually agreed upon purpose for meeting together must be identified. This first goal may be an agreement to work on the presenting problem, to listen to the client in order to help the client gain some insight and a better understanding of himself or herself, or to explore the client's rather diffuse concerns in order to develop an understanding of the client and thereby identify a more precise counseling goal.

STRUCTURING THE RELATIONSHIP.   All counseling relationships need to be structured in order to provide the client with a sense of clarification, direction, and understanding of the process. Initially, a client may not have a clear picture of what to expect or how changes are likely to occur, and he or she may be uncertain of the logistic and procedural aspects of this process. As the counselor you should develop a mutual understanding and agreement between yourself and the client regarding the dynamics and methodology of the process; the logistical, pragmatic, and procedural issues involved; and the client's personal or consumer-oriented concerns (Day & Sparacio, 1980).

To develop an understanding of the dynamics of the counseling process you may need to discuss how counseling works, the nature of the interaction, the responsibilities of both yourself and the client, and information about how positive results can be achieved. For example, as a college counselor you may provide this structure to a student who is undecided about his major by stating, "My job is to help you think about the issues that are involved in making this decision. You are responsible for obtaining any additional information that you need, judging its relative importance, and weighing it in light of your own personal background and values."

As the counselor you should be sure that the arrangements are clear for all the important logistical and practical details, such as time, including time of meeting, session length, frequency of contact, and how long the process may last; location, including address, building, and room number; arrangements for absences, cancellation, and emergency situations; and other details such as costs. For example, as a rehabilitation counselor you may say to your client, "John, we will meet every Thursday afternoon from three until three forty-five, in room 415 in the rehab center. If for some reason you cannot make it, please call the office and leave a message for me. We can reschedule our session for another time that week."

The third aspect of structuring is to make sure that there is an agreement about those items that deal with personal concerns of the client, such as the confidential nature of the relationship, arrangements for recording the sessions, the qualifications of the counselor, and what the client's responsibilities are in the process. For example, as a counselor working in a community agency you may say, "Anything you say in these sessions will be kept confidential. Our sessions will be taped so that I can review the highlights of your case with my supervisor."

The amount of structure that is provided differs from case to case and will depend on the needs of the client, the environmental setting, the type of problem presented, and your theoretical orientation. Frequently counselors structure the relationship in general terms initially and provide additional structure as the need develops. Too much structuring can increase a client's anxiety rather than lower it, and too little can create an erroneous perception of what the counseling process is all about.

MAINTAINING ETHICAL STANDARDS.   As the counselor you need to exemplify appropriate ethical behavior throughout the counseling process. At the initial stage of counseling, the major ethical concerns involve confidentiality, informed consent, and the appropriate representation of your skills and credentials. These issues should be addressed when you structure the relationship, but you will need to continue to be concerned about ethical standards throughout the entire counseling process.

The Exploratory Stage

During the second transitional stage you should focus on assisting the client in exploring, perceiving, analyzing, and understanding the parameters of self and his or her problem. To gain an understanding of the client and his or her internal frame of reference, it is most helpful to address the following four major tasks at this stage:

- explore the scope or depth and breadth of the client's presenting problem;
- analyze the client's degree of functioning in several major dimensions of the client's life;
- understand the historical and idiosyncratic ways that this particular client has changed and the amount of resistance that presently exists toward change; and
- identify the client's internal strengths and the external resources available to the client.

At this stage you may help your client move from a surface awareness of his or her concern to a recognition of any underlying issues that need to be addressed.

THE SCOPE OF THE PROBLEM.   The major task at this stage is to explore and understand the reasons the client came for help. To facilitate this exploratory process both you and the client need to clarify whether the presenting problem is the major or real problem; identify the context of the environmental situation where the problem occurs; evaluate the nature, severity, and duration of the problem; assess the consequences that this particular problem has caused the client and significant others; and estimate the effects that any change would have on both the client and significant others in the client's life. This exploratory process is often best done by actively listening to the client, clarifying issues, and supporting and encouraging the client as a person of value. Open-ended inquiries may be used, but care must be taken to avoid creating a dependency relationship or an interrogative atmosphere.

THE CLIENT'S DEGREE OF FUNCTIONING.   To understand the client's internal frame of reference, it is usually helpful to examine the client's degree of functioning in several major aspects of life. As the exploratory process unfolds, you may form a professional judgment about whether the client is progressing through the developmental life stages with a minimum of problems; meeting his or her physiological, psychological, and social needs in effective ways; functioning at an appropriate cognitive level; relating in positive ways to significant others; coping and adjusting well to the conflicts, frustrations, and other thwarting conditions encountered in life; and manifesting appropriate behavioral patterns. An examination of these dimensions will provide further insight into the client and increase both the client's and your awareness of the ability of the client to function in a variety of dimensions of life.

HISTORICAL PATTERNS.   As the counselor you may want to help a client explore and understand how he or she has historically dealt with problems similar to the presenting one. For example, it may help a particular client for you to know if the client has discussed important problems with significant others, sought new factual information to clarify issues, tried out new experiences or roles, changed his or her environment, or took chances when the future was uncertain. This exploratory process can reveal the client's resistance to handling life issues or his or her ability to deal with the uncertainty of the problem-solving process. During the exploration process, you will begin to formulate some tentative plans about the intervention strategy you may want to employ with the client.

STRENGTHS AND RESOURCES.   The client's internal strengths and the source and kinds of resources available to the client may also be explored during this stage. To understand the scope of the client's strengths, it is helpful to ascertain the client's sense of ownership for the problem, sense of responsibility for resolving the issue, awareness of strengths and limitations, and knowledge of available external resources. Helping clients obtain this awareness and a sense of responsibility is essential in counseling. The way you treat and respond to your clients will affect their willingness to take responsibility and ownership for their own thoughts, feelings, and actions. If clients experience you as a caring, trusting, and concerned helper, they will slowly become less defensive and more open to self-exploration and self-understanding. Helping clients uncover issues they may not be aware of brings about clarification of the problem and often suggests an intervention strategy.

The Decision-Making Stage

The third transitional stage is intermediate between the exploration of a client's concerns and the application of a particular intervention plan. Two interrelated tasks need to be accomplished at this stage: the goals of the counseling process must be mutually agreed upon by you and the client, and a decision should be made by you regarding the particular intervention strategy that will be used. The initial goals may be redefined or revised on the basis of the understanding reached in the first two stages of counseling. It is critical that the goals be mutually agreed upon because progress will be most unlikely if you are working on one concern while the client has a need to work on something quite different. The extent to which this agreement should be explicit varies from one approach to another. If you plan to use strategies based on a person-centered approach, you could accept an implicit agreement; whereas if you plan to employ a behavioral approach, you would want to have a very explicit agreement.

Deciding on a goal is not always a clear-cut process. Some clients present a multitude of issues that may impede their ability to function effectively. In these cases it is strongly recommended that as a beginning counselor you give serious consideration to working on one concern that is causing major discomfort and yet is a goal that can be reached within a reasonable length of time. Reaching a decision on a counseling goal and the type of intervention is a task that cannot be accomplished effectively unless the tasks of the prior stages have been met. There are a number of client, counselor, and environmental variables that will influence the specificity of these counseling goals and the intervention strategy used.

CLIENT VARIABLES.   The client characteristics that strongly influence the counseling goal and the intervention approach include:

the kind of problem—for example, does the client have poor interpersonal skills or does he or she have poor decision-making skills?

the historical and idiosyncratic pattern employed to solve problems and resolve issues—for example, does the client typically let others make the decisions or does he or she have the habit of investigating all possibilities so thoroughly as to become paralyzed in the process?

demographic characteristics—for example, does the client's age, gender, or income level have a bearing on the college and the financial aid that he or she is eligible for?

personality characteristics—for example, the dependent client will have to be handled somewhat differently from the independent client.

COUNSELOR VARIABLES. The counselor variables that strongly influence the choice of the counseling intervention include the counselor's:

knowledge of the cognate area—for example, theories and research in motivation, cognition, relationship, adjustment, and personality of behavior;

knowledge and experiences in using various counseling approaches—for example, person-centered, cognitive restructuring, or behavioral; and

level and skill in communicating appropriate high-level responses—for example, helping skills, verbal responses, and role-communication skills.

ENVIRONMENTAL VARIABLES. The setting (school, college, vocational-rehabilitation agency, or private practice) where the counseling takes place may have a moderate to strong influence on the appropriateness of a particular goal or intervention strategy. Counselors who work in institutional settings often delimit their goals in order to help the client function more effectively within that setting. For example, a school counselor working with a child who has an alcoholic parent must have an immediate goal of helping the student function more effectively at school despite the family problem.

The Working Stage

The fourth counseling stage is the application of an appropriate intervention strategy. The major task at this stage is helping the client resolve his or her concern and learning to function more effectively. This may require you to provide emotional support, encouragement, and reinforcement of newly gained insights. You may want to use one particular strategy or a multivariate treatment strategy. The strategies employed in counseling can be classified as interventions that emphasize improvement in the client's level of mental functioning, sense of well-being or emotional state, or ability to behave more appropriately. Thus treatment strategies may be cognitively focused, affectively focused, or performance focused. *Cognitively focused* interventions should be considered when you believe that the client needs assistance to obtain or retain factual information, when the client needs help making decisions, or when the client reveals faulty deductive or inductive thinking processes. *Affectively focused* interventions should be considered when you conclude that the client reveals inadequate feelings of self-worth, poor acceptance of others, and minimal skills in dealing with his or her own attitudes, beliefs, emotions, or values. *Performance focused* interventions may be appropriate when the client's behavioral repertoire is limiting his or her functioning. More often than not, counselors will employ a multimodal approach using strategies from all three domains. Intervention strategies based on these three modalities are explained more fully in Chapters 4, 5, and 6.

In the working or intervention stage of counseling, counselors often use several resources to assist clients in resolving their concerns. These resources may be significant others in a client's life, community resources, or published materials.

SIGNIFICANT OTHERS. Frequently an appropriate intervention strategy involves significant other persons in a client's life. The client may be required to interact in a different way than previously with a parent, spouse, child, teacher, or employer. When this is probable, your treatment approach must give careful attention to this aspect of the client's life. At times, you may want to involve these significant other persons in the intervention strategy. When this is desirable, appropriate consultation with the concerned parties should be conducted.

COMMUNITY RESOURCES. Community resources are used in the counseling process for three distinct purposes. First, clients may be referred to another person or agency for complete treatment if you realize that the client has a problem that can best be resolved through the assistance of another professional person or of a particular group or association. Examples of such referrals include optometrists, physicians, psychologists, and speech and hearing specialists. Referrals may also be made to a specialized agency, such as the Association for the Help of Retarded Children or the United Cerebral Palsy Association. Second, clients may be referred for concurrent treatment if you recognize that other professional help is needed by the client as part of or perhaps parallel to the counselor's intervention. For example, as a school counselor you may refer a student who has an alcoholic parent to Alateen, a support group for children of alcoholics. Finally, community resources are also used to enable clients to obtain important information necessary to advance the counseling process. Examples of this type of referral include the following:

- you are a vocational counselor and your client is concerned with career choice, so you want that client to interview a person who is working in a particular occupation;
- you are a school counselor and your client is in the process of selecting a college, so you want your client to visit the colleges that he or she is considering; or
- you are a vocational rehabilitation counselor and your client may have recently suffered some disabling event and needs to be directed to a particular agency to learn a new trade.

PUBLISHED MATERIALS. Published materials and references are valuable in the treatment phase of counseling to help the client obtain information that will enhance the treatment. For example, a student who is interested in becoming a dental hygienist may be directed to the *Occupational Outlook Handbook* to learn more about the occupation.

The Termination Stage

During the fifth and final transitional counseling stage, the termination process occurs. This is an extremely important period in which you need to focus on accomplishing three interrelated tasks. First, progress made should be summarized and evaluated. Second, other issues that require attention at this time should be brought forward. And third, methods to foster client growth after the counseling process terminates need to be established. When these tasks are handled effectively, the counseling process is successfully completed. When these tasks are not dealt with, the process is truncated and the important growth that has occurred is curtailed.

To evaluate the counseling process, you and the client should determine whether the desired goals were met. The major responsibility for accomplishing this first task should be placed on the client (Ward, 1984). The client may be asked to prepare a progress report indicating how the counseling goals were reached. This may entail having the client state how he or she has changed, what new learning has occurred, or how he or she is better able to deal with specific situations or significant other persons. To consolidate the progress you ought to review the client's progress report carefully and then verbally summarize and review what happened and why it happened. This consolidation will serve to reinforce attainment of the goals.

Determining the client's preparedness to terminate requires an assessment of the client's overall level of functioning and whether or not other client concerns or unresolved issues need to be addressed at this time. One issue that frequently needs attention is dependency. There is often a natural tendency to maintain the bond that was established in the counseling process, and the dependent client will manifest this tendency to a high degree. These dependency feelings must be dealt with before the termination process can occur. Clearly, not all client issues can be resolved; clients will never function fully in all aspects of their lives. But once the agreed-upon goals have been met, unless another goal that is mutually agreed to is selected and the counseling stages are recycled, the counseling process should move toward closure.

The final task in the termination process is helping the client develop a systematic method to ensure that the growth and change process will continue. This ordinarily includes arrangements for periodic follow-up sessions, as well as developing a self-monitoring plan and rehearsing ways this plan will be implemented. Self-monitoring plans and follow-up efforts serve to improve clients' self-confidence and provide a necessary support system.

Termination is often a hard step, but when the tasks of this stage have been accomplished the client should be ready to leave. During the final session the client should be informed that the relationship is not being ended but the meetings are being adjourned; the client should be encouraged to return whenever he or she needs further assistance.

RELUCTANT AND RESISTANT CLIENTS IN THE COUNSELING PROCESS

Many clients come to see a counselor on a voluntary basis. These clients recognize that they have an unresolved problem—they have the motivation to obtain professional assistance, and they have made a commitment to involve themselves in the change process. Other clients, however, are referred to a counselor involuntarily. They are compelled to enter the process by some pressure outside themselves. Dyer and Vriend (1975) maintain that most of the clients seen in institutional settings are involuntary. This section confronts some of the basic issues counselors face in dealing with both involuntary and voluntary clients.

The Reluctant and Uncommitted Client

The reluctant client is one who is unmotivated to seek help; if it was left up to this person, he or she would never go talk to a counselor. The counseling process is not seen as a reasonable or realistic approach. When a reluctant client does appear and starts this process, the likelihood that the counseling process will be incomplete is strong unless you take some effort to prevent this. In school and other institutional settings, involuntary clients include students referred for poor classroom achievement, disciplinary concerns, and other maladaptive or ill-advised behaviors. In rehabilitation and other community agencies, uncommitted clients may be referred by a concerned or overwhelmed parent, another agency, the probation system, or the courts.

These uncommitted clients may be unwilling to become committed for a variety of reasons. Many see the counseling process as an affront to their own self-concepts. They believe that the way they are functioning is okay; any action showing a willingness to change or to seek help is an admittance of their own weakness, a sign of failure. Others see the counselor as part of the system that they are already at odds with—the very authorities who have caused the client difficulty are now using one of their colleagues to set the client straight! Still others are reluctant to change because their ill-advised behavior may have given them some status with their peer group. Thus any change in behavior will necessarily result in a change of status, and that may be quite undesirable. Others see the counselor as a person who is trying to control their lives; therefore the clients' attempts to be independent are threatened.

Client reluctance is manifested in many ways. A principal one is the silent treatment. Reluctant clients may refuse to discuss anything, and when they do they nod, shrug their shoulders, or give short answers to any and all questions. Beginning counselors who are not careful and prepared to deal with reluctant clients may intensify their questioning, creating an interrogative atmosphere; they get nowhere, and their reluctant clients remain minimal communicators.

Another reluctant group consists of the avoiders, who are seemingly quite agreeable and compliant. They are willing to talk about anything and everything but the real issues. Their avoidance of work on the important topics is signaled by their loquaciousness, silly actions, or willingness to work only on small inconsequential concerns. Reluctance is shown in a third way by clients who have excuses for everything they do. Their defense mechanisms are powerful and serve as a protective shield. A final way that this unwillingness is displayed is by hostile actions. The angry client appears to have no tolerance for any institution or system and vents this hostility at the least provocation.

Handling the uncommitted and involuntary client is difficult for any counselor and extremely frustrating for the beginning counselor. Trying to deal with this type of client can cause counselors to blame themselves, feel a sense of personal failure, and develop a lower professional self-regard. Furthermore, when little progress is made in counseling, there is a very strong possibility that a counselor will reinforce the client's reluctance. This behavior is reinforced by being impatient; ignoring the client's signals; getting upset and directing irritation at the client, to whom one should be sending signals of positive regard; and ultimately giving up and refusing to do any further work with the unwilling client.

To deal with an uncommitted client, you should first be aware of the need to establish realistic expectations for the client and the counseling session. Asking yourself the following questions can be helpful:

- Is the client coming voluntarily or was the client referred by a third party?
- Did you expect to accomplish a great deal or did you expect to move rather slowly?
- Did you expect a compliant, easygoing individual or a person who was defending his or her own concept of self?

Formulating realistic expectations will help you prevent feelings of frustration and positively influence the counseling process. Second, you need to continue communicating a warm, deep respect for the client. Unless the client feels that you are on his or her side, the reluctance will be maintained. Third, the client's feeling of self and self-expression must be dealt with in the counseling process. The reluctant client's major interest is the self. Therefore, almost any technique that enhances the client's self-understanding will serve to reduce or lower the reluctance.

Reluctance should be dealt with as it comes up in counseling. As in dealing with any other deeply felt issue, you will need to help the client gain an understanding of the reluctance, become aware of this reluctance, and take steps to deal with it in an effective manner. You must be in touch with the client, acknowledge his or her feelings, and possibly interpret the client's behaviors. This interpretation could be done in ways similar to the following: "Silence helps you feel more comfortable, and discussing issues that are of concern to you appears to be too painful"; "Your reluctance to discuss this important topic may be saying to both of us that you are quite comfortable in staying where you are"; or "Your not wanting to discuss the important issues appears to be counter-productive." Counselors can anticipate how they will work with reluctant clients by simulating these cases before they occur using role-playing and role-reversal techniques.

Strategies that go beyond those used within the typical dyadic counseling process are frequently very helpful when working with reluctant clients. Many effective strategies have been developed to work with clients who manifest ill-advised behaviors (Brown, Pryzwansky, & Schulte, 1987). In order to help clients overcome some self-defeating behaviors, you may need to work with signifi-cant other persons in the client's world. Modification of the client's environ-ment may be helpful, and other resource persons may be useful.

The Resistant But Committed Client

The resistant client is typically one who volunteered to come for help, entered into the relationship, and became at least superficially involved in the counseling process but is unwilling to change his or her feelings, thoughts, or overt behaviors. This resistance to reach a decision, to recognize symptoms, or to give up self-defeating activities is counterproductive for the client and quite frustrating for the counselor. Some believe that resistance is pervasive throughout all counseling and therefore present to some extent with every client (Peterson & Nisenholz, 1987).

Learning about oneself and taking the steps to change self-defeating thoughts, feelings, or behaviors can be threatening. This resistance often is not a conscious attempt to thwart the counseling process, but it is a real barrier to progress. The resistance to avoid dealing with the issues and doing any hard work takes many forms. Some clients are silent, some appear very tired and listless, some act quite forgetful or evasive, some hide behind a barrage of words, and others become quite defensive or argumentative.

As a counselor you can learn to deal with this resistance when clients manifest it. First, you must anticipate that it may occur and not become anxious or defensive. Second, you need to continue showing a warm, caring, and con-cerned interest in your client. Third, try to understand what is causing the resistance: is there an overt or covert payoff for the client if he or she does not change? And finally, deal with the resistance in a constructive and helpful way. There are several alternative ways to handle this resistance positively. If the resistance is relatively mild, it may be best to ignore it and move ahead. If the resistance is more serious, you may wish to downplay it and direct the client to move ahead with the next step in the counseling process. If the client is highly anxious and fearful of moving forward, humor and diversionary tactics may be helpful. Gladding (1988) and Roloff and Miller (1980) suggest that it is helpful to have clients take steps to do something quite minor and then something quite major, or to ask them to do something impossible and then something quite reasonable. Either way, some action is accomplished and a resistance block may be broken. Another way to handle this situation is to help the client understand that resistance is present and use the client's awareness of it as part of the counseling process.

SOURCES OF INFORMATION ABOUT THE CLIENT

The purpose of this section is to describe several important ways that counselors typically obtain information about clients. Although the primary means of obtaining information about the client is through the verbal interchange between the client and the counselor, information is often obtained before and during the counseling process in a variety of other ways. The most common methods are (1) prior personal knowledge of the client, (2) reports from significant others, (3) life-history questionnaires, (4) environmental or situational observations, and (5) psychometric data. Each of these data-collection techniques and resources is discussed briefly in the following sections.

PERSONAL KNOWLEDGE OF THE CLIENT.   Counselors sometimes know their clients before beginning the counseling process. Many counselors work in institutional settings such as schools, colleges, or residential treatment programs and have previously met clients and observed them in a variety of situations within the institutional setting. This prior knowledge can minimize the amount of time spent on the relationship-building stage and often facilitates the exploratory stage. As a counselor, you should acknowledge this previous information early in the counseling process in order to ensure that the interpretations of these previous observations are valid and known to both the client and the counselor.

REPORTS FROM SIGNIFICANT OTHERS.   Reports about the client may be written or verbal, brief or extensive, and may come from people who have known the client either in a professional or in a personal relationship. One of the more common forms is a report written by another professional person. These reports will vary in the type and extent of information supplied. One example is a minimally informative school anecdotal record, such as "John was quiet in the sixth grade." Another is an informative referral from a teacher, such as "John is failing in sixth grade. He can respond well to verbal questions; however, on written tasks he responds poorly. I suspect his auditory learning style is good but his visual one is poor. Can you pursue this matter with John and/or his parents?" Another example is a rather extensive psychological report concerning the client's past behavior and the prognosis for his or her future.

LIFE-HISTORY QUESTIONNAIRES.   Many counseling agencies require new clients to complete a life-history questionnaire prior to the initial or intake interview. The purpose of this questionnaire is to obtain a comprehensive picture of the client's background and facilitate the exploratory process. These questionnaires probe into the client's background and often inquire about the following areas:

- General information: name, age, sex, height, weight
- Residential data: address, phone number, living environment
- Familial data: marital status, children, parents, siblings
- Educational data: schools attended, years of attendance, major
- Occupational history: jobs held and companies worked for
- Avocational interests: hobbies, sports, civic activities
- History of physical health: previous illnesses and accidents, present health status, medication currently taken
- The presenting problem: its history and any previous counseling experience

A life-history questionnaire may be relatively short or quite long and extensive depending on the philosophy of the agency.

ENVIRONMENTAL OR SITUATIONAL OBSERVATIONS. To gain a better understanding of their clients, counselors often see their clients in the environment that is related to the clients' concerns. This is particularly true for counselors who work in institutional settings as well as for those who work in marital counseling. For example, counselors visit the classroom to see how a child or student is functioning with his or her peers in that room; rehabilitation counselors visit the sheltered workshop to see how a rehabilitation plan is working; and marriage counselors frequently see a couple together and also schedule sessions when the entire family can be present.

PSYCHOMETRIC DATA. Standardized tests provide an objective and standardized methodology that counselors in various settings often use. These instruments are employed for four different reasons.

First, they are used to identify individuals who can benefit from the counseling process. Counselors who work in institutional settings are responsible for counseling large numbers of individuals, so they use various screening inventories or checklists for identifying clients who can most benefit from counseling.

Second, counselors use psychological instruments to help clients see an objective picture of their strengths and limitations. For example, school and rehabilitation counselors administer and interpret scholastic-aptitude tests, achievement tests, and interest inventories so that their clients can compare themselves to others objectively and gain further insight into their educational and vocational development.

Third, counselors use tests to measure the growth of individuals or groups of individuals over time or to see if a particular counseling intervention facilitated a particular growth. For example, a school counselor may give an interest inventory to a student before vocational counseling and again after the counseling to see if the individual's interests have changed as a result of the counseling process.

Fourth, counselors sometimes use tests to gain knowledge about the counseling process by comparing the relative efficacy of two or more counseling approaches or interventions. For example, a vocational counselor may wish to research whether bibliotherapy, reading about occupations, is a more effective treatment than seeing role models or hearing lectures by experts. To investigate this question, the counselor would set up an appropriate experimental study to find the answer.

SUMMARY

This chapter has outlined the characteristics of the counseling process as it progresses through the five transitional developmental stages. The chapter also provided some important concepts for working with resistant and reluctant clients. In addition, five different sources that counselors often use to obtain information about their clients were briefly presented. After reading and studying this chapter you should be able to identify the five transitional counseling stages and discuss the major characteristics of each stage. You should be able to discuss the differences between reluctant and resistant clients. And you should be able to describe the advantages and limitations of five different sources of information about clients.

REFERENCES

Baruth, L. G., & Huber, C. H. (1985). *Counseling and psychotherapy: Theoretical analysis and skills applications.* Columbus, OH: Merrill.

Brammer, L. M. (1988). *The helping relationship* (4th ed.). Englewood Cliffs, NJ: Prentice-Hall.

Brown, D., Pryzwansky, W. B., & Schulte, A. C. (1987). *Psychological consultation: Introduction to theory and practice.* Boston: Allyn & Bacon.

Day, R. W., & Sparacio, R. T. (1980). Structuring the counseling process. *Personnel and Guidance Journal, 59,* 246–249.

Dyer, W. W., & Vriend, J. (1975). *Counseling techniques that work.* Washington, DC: APGA Press.

Egan, G. (1990). *The skilled helper: A systematic approach to effective helping* (4th ed.). Pacific Grove, CA: Brooks/Cole.

Gladding, S. T. (1988). *Counseling: A comprehensive profession.* Columbus, OH: Merrill.

Hansen, J. C., Stevic, R. R., & Warner, R. W., Jr. (1986). *Counseling: Theory and process* (4th ed.). Boston: Allyn & Bacon.

Ivey, A. E., Ivey, M. B., & Simek-Downing, L. (1987). *Counseling and psychotherapy* (2nd ed.). Englewood Cliffs, NJ: Prentice-Hall.

Lerner, R. M. (1986). *Concepts and theories of human development* (2nd ed.). New York: McGraw-Hill.

Okun, B. F. (1987). *Effective helping* (3rd ed.). Pacific Grove, CA: Books/Cole.

Patterson, L. E., & Eisenberg, S. (1983). *The counseling process* (3rd ed.). Boston: Houghton Mifflin.

Peterson, J. V., & Nisenholz, B. (1987). *Orientation to counseling.* Boston: Allyn & Bacon.

Roloff, M. E., & Miller, G. R. (Eds.). (1980). *Persuasion: New directions in theory and research.* Beverly Hills, CA: Sage Publications.

Turiel, E. (1969). Developmental processes in the child's moral thinking. In R. H. Mussen, J. Langer, & M. Covington (Eds.), *Trends and issues in developmental psychology* (pp. 92–133). New York: Holt, Rinehart & Winston.

Ward, D. E. (1984). Termination of individual counseling: Concepts and strategies. *Journal of Counseling and Development, 63,* 21–25.

How Counselors Help Individuals Change: Counseling Intervention Strategies

The beginning student in the field of counseling is often bewildered by the sheer number and the conceptual diversity of approaches to counseling. This confusion can be resolved in one of three ways. One way is to focus on one and only one theoretical approach and practice it until one becomes an expert in using that approach. The counselor-in-training who follows this method will become identified with a particular theory such as a person-centered counselor, a trait-factor counselor, or a behavioral counselor. A second way is to focus on one theory or approach, practice it until some basic competencies are achieved, and then go on to develop some skills using another theory. Thus, one can learn to counsel by mimicking Rogers, then perhaps Perls or Ellis or Lazarus, until one becomes familiar with a variety of approaches. In this training modality, the counselor trainee is faced with the responsibility of mastering a variety of schools and learning to employ a particular model for a particular client need. This is the menu approach to the counseling process. A third way is to formulate a unified conceptual model that slowly but systematically integrates ideas and techniques from all the major approaches as one develops one's own counseling style and modality.

This section of the text discusses contemporary approaches to integrating various counseling approaches; provides a rationale for working with clients from the cognitive, the affective, and the behavioral modalities; and outlines important counseling intervention strategies based on these modalities. Chapter 4 delineates several important cognitively focused interventions. Chapter 5 is devoted to a discussion of the major affectively focused interventions. And Chapter 6 describes the counseling interventions based upon behaviorally focused modalities.

CONTEMPORARY THEORETICAL DEVELOPMENTS

There is ample evidence in the literature that various counseling approaches have merit and that one theory or approach is not superior to all other theories or approaches (Cormier & Cormier, 1991; Frank, 1982; Strupp, 1982). Most help-ers use a variety of approaches (Norcross & Prochaska, 1988). There are several indications in the counseling field that there is a trend to develop a pantheoreti-cal, or more universal or united, approach to the helping process. First, increas-ing emphasis has been placed on teaching beginning counselors a repertoire of communication skills that are useful in different stages of counseling and in a variety of approaches to the counseling process (Danish, D'Augelli, Hauer, & Conter, 1980; Doyle, 1982; Egan, 1989; Hill, 1978; Ivey, Ivey, & Simek-Downing, 1987; Kagan, 1972). Second, there appears to be a greater focus on learning theory and more awareness of the role and importance of the client's motivation, self-expectations, and self-attributions as essential ingredients to the process of change (Frank, 1982; Kanfer & Gaelick, 1986; Linehan & O'Toole, 1982). Third, a consensus is emerging that counseling is based on developing a sound interper-sonal relationship and employing sound technical strategic interventions (Frank, 1982; Strupp, 1982). Fourth, efforts have been made to organize the many theoretical approaches into meaningful taxonomies of intervention strategies (Frey & Raming, 1979; Hutchins, 1979; L'Abate, 1981). Finally, there is considerable discussion in the literature about the feasibility of developing an integrated ap-proach to the helping process (Erskine & Moursund, 1988; Goldfried, 1982; Goldfried & Wachtel, 1987; Patterson, 1985; Prochaska & Norcross, 1986).

CHANGE IS A NATURAL OCCURRENCE

Clients do not change in mysterious ways. Change is a dynamic and normal part of our everyday existence. Individuals grow, develop, and modify their behavior in a variety of ways. The following examples illustrate twelve ways in which this can and does occur.

1. *Receiving direct instructions.* People respond to direct instructions and factual information. For example, the Postal Service reminds us to mail our let-ters early at Christmastime with the admonition "Mail Early!"; and on the ski slope, the ski instructor is heard saying, "Keep your weight on the downhill ski."

2. *Learning how to make choices.* Choosing between and among alter-native courses of action is important in our everyday lives. Individuals must learn to weigh the alternatives, consider the advantages and disadvantages of doing a particular thing, and recognize and accept some degree of risk in every choice. Examples of some major choices that affect our behavior include the college sophomore asking "What shall I major in?" and the newlywed couple ponder-ing "What neighborhood should we live in?"

3. *Learning how to think deductively.* Thoughts and beliefs can and do influence our emotions and our daily activities. Consequently, thoughts that are based on a faulty reasoning process lead to faulty conclusions and perceptions; thus, we can change by learning to think more logically. For example, the state-ment "I am no good" may be based on the two premises "I have difficulty do-ing something" and "People who have difficulty doing something are no good";

thus, "I am no good." In this case, the error in the second premise needs to be pointed out and emphasized so that the individual can reason more logically.

4. *Discussing an issue with someone you respect.* A discussion of one's thoughts, feelings, or frustrations can often help a person change. For example, when a close friend is very angry or is feeling a deep loss, we can provide the atmosphere for the person to air his or her concern and thus talk out the anger or the depression.

5. *Gaining more insight.* Helping a person look at himself or herself in a way that promotes self-knowledge can influence how a person thinks, feels, or acts. For example, the student who becomes aware of the fact that certain actions have a negative effect on others should gain some feeling that his or her behavior ought to change.

6. *Enhancing one's self-esteem.* A person who has more self-confidence and positive feelings about himself or herself is more likely to behave assertively than an individual who has a poorer self-image. For example, the young man who has a good self-image is more likely to have positive interpersonal relationships than is the individual who has a poorer self-image.

7. *Watching another person.* When an individual observes another person perform, that individual is likely to replicate that behavior. For example, tennis students carefully watch their tennis coach, and counselors-in-training want to see an expert, such as their instructor, perform.

8. *Drilling and practicing.* Performing a new activity over and over again until it is part of our repertoire is one way to ensure that a particular action is likely to be repeated. For example, an individual who is unsure of his or her performance in a potential job interview will practice or role-play an imaginary interview until more confidence is gained.

9. *Changing the environment.* Individuals can increase the probability that they will act in a certain way by changing the conditions under which the activity takes place. For example, when an individual is studying for a final examination, he or she probably wants to remove distracting cues by studying in a quiet place. Another example is the person who wants to stop smoking—he or she does not buy cigarettes and does not associate with people who smoke.

10. *Learning how to reduce anxiety.* Some individuals are unable to do certain things because a great deal of anxiety is associated with that particular situation. By learning how to reduce the anxiety, they can be freer to act. For example, a student can be taught to relax before taking an examination, or an employee can be taught to think of pleasant scenes when he or she has to do something anxiety-provoking at work.

11. *Receiving positive feedback.* When an individual acts in a positive way and receives praise for that endeavor, that behavior is likely to be repeated in the future. For example, the counselor-in-training who is told by his or her peers and supervisor that a particular skill was performed well will be apt to use that skill again under similar circumstances.

12. *Receiving negative feedback.* When a person does something that causes others to frown at him or look down on her, then that person is likely to avoid acting that way in the future. For example, the young child who picks

up an ashtray may be admonished by his parents with the sharp words "No, No!" or the student who doesn't hand in his or her homework on time may receive a demerit. The vocal rebuff and the demerit are forms of punishment that are used to decrease the likelihood of those behaviors. Although negative feedback can be effective, too much negative feedback can enhance feelings of inadequacy, so it should be used very sparingly.

THE THREE DOMAINS OF FUNCTIONING

Human beings are in a dynamic interactive relationship between themselves and their environment. Individuals strive to maintain their equilibrium throughout experiences that bring about constant changes and adaptations. Everyone learns new things and begins to think, feel, believe, and behave differently as a result of interactions with societal or environmental influences. Because individuals are unique, the influence that societal forces have varies with their patterns of interactions and their striving for equilibrium. Individuals change in a variety of ways and for unique reasons.

Historically, counseling strategies or interventions were classified using two-dimensional and linear formats. In the fifties and early sixties the directive versus nondirective dichotomous scale was in vogue. This was replaced by the rational-affective framework discussed by Barkley (1968), Patterson (1986), and Shertzer and Stone (1980) and employed by Frey (1972) in the analytical goal and process taxonomy.

Hutchins (1979) and L'Abate (1981) have maintained that counseling theories need to be viewed from a three-dimensional construct. Hutchins used the terms *thinking, feeling,* and *acting* for his *T-F-A* model, which classifies theories into various combinations of these modalities depending on the primary counsel-ing intervention. Thus Rogers is seen as an F-A-T (feeling-acting-thinking) or F-T-A (feeling-thinking-acting) counselor, whereas Ellis is considered to be a T-A-F (thinking-acting-feeling) or T-F-A (thinking-feeling-acting) interventionist. L'Abate presented a very similar eclectic model. It uses the terms *emotionality, rational-ity,* and *activity,* and hence is labeled the *E-R-A* model. This model classifies theorists using a process and goal format and comes to slightly different con-clusions on the placement of various theoreticians.

Strupp (1982) pointed out that the therapeutic process is essentially a tutorial, collaborative, educational one and that success in therapy should be demonstrable in one or more of three functioning domains—that is, in the client's mental functioning, sense of emotional well-being, and social functioning or behavioral activities.

The three major interactive functioning domains should not be used only to classify theorists and to measure the success of the counseling process. They can and should be used in understanding a client's functioning status and sug-gest a meaningful taxonomy for classification of counseling interventions. Thus, the specific strategies or technical interventions that counselors employ may be considered as cognitively focused, affectively focused, or performance focused. These three domains are illustrated in the diagram that follows.

The major domains of functioning

It is important to note and realize that changes in one of these domains can have an effect on one or both of the other domains. Therefore, when we alter the way we think about something, we normally change the way we feel about it and the way we behave toward it. Similarly, when we change the way we feel about somebody, we usually alter the way we think about and act toward that person. The same is true for the third domain; that is, when we change the way we behave, modifications occur in our thinking and feeling domains as well.

THE COUNSELOR'S ROLE

The counselor's role in helping individuals change is to facilitate these natural avenues. Counselors, by definition of this role, interject themselves into their clients' interactive relationship with their environment and attempt to facilitate natural growth and the dynamic process. The major goals of the counseling process are to assist clients in overcoming impediments to change and becoming more self-directed, and to facilitate both psychological independence and interdependence so that clients are free to act for their own good as well as the good of society.

In the process of understanding these technical interventions or counseling strategies, it is essential that counselors-in-training understand that three important concepts underlie all interventions: the professional working relationship between the counselor and the client; the counselor's linguistic and communication skills; and the client's motivation, expectations, and self-attributions. Furthermore, it must be noted that interventions are rarely used alone; more often counselors use multimodal approaches that use strategies from the cognitive and affective as well as the performance domains.

Specific intervention techniques based on cognitive, affective, and performance concepts are outlined in the following three chapters.

REFERENCES

Barkley, J. (1968). Counseling and philosophy: A theoretical exposition. In B. Shertzer & S. C. Stone (Eds.), *Guidance Monograph Series, Series II: Counseling.* Boston: Houghton Mifflin.

Cormier, W., & Cormier, L. S. (1991). *Interviewing strategies for helpers: Fundamental skills and cognitive behavioral interventions* (3rd ed.). Pacific Grove, CA: Brooks/Cole.

Danish, S. L., D'Augelli, A. R., Hauer, A. L., & Conter, J. J. (1980). *Helping skills: A basic training program* (2nd ed.) New York: Human Sciences Press.

Doyle, R. E. (1982). The counselor's role communication skills: The roles counselor's play. *Counselor Education and Supervision, 22,* 123–131.

Egan, G. (1989). *The skilled helper* (4th ed.). Pacific Grove, CA: Brooks/Cole.

Erskine, R., & Moursund, J. (1988). *Integrative psychotherapy in action.* Newbury Park, CA: Sage Publications.

Frank, J. D. (1982). Therapeutic components shared by psychologists. In J. H. Harvey & M. M. Parks (Eds.), *Psychotherapy research and behavior change* (Vol. 1, pp. 5–37). Washington, DC: American Psychological Association.

Frey, D. H. (1972). Conceptualizing counseling theories: A content analysis of process and goal statements. *Counselor Education and Supervision, 11,* 243–250.

Frey, D. H., & Raming, H. E. (1974). A taxonomy of counseling goals and methods. *Personnel and Guidance Journal, 58,* 26–33.

Goldfried, M. R. (Ed.). (1982). *Converging themes in psychotherapy: Trends in psychodynamic, humanistic, and behavioral practice.* New York: Springer.

Goldfried, M. R., & Wachtel, P. L. (1987). Clinical and conceptual issues in psychotherapy integration: A dialogue. *Journal of Integrative and Eclectic Psychotherapy, 6,* 131–144.

Hill, C. E. (1978). Development of a counselor verbal response category system. *Journal of Counseling Psychology, 25,* 461–468.

Hutchins, E. E. (1979). Systematic counseling: The T-F-A model for counselor intervention. *Personnel and Guidance Journal, 57,* 529–531.

Ivey, A. E., Ivey, M. B., & Simek-Downing, L. (1987). *Counseling and psychotherapy: Skills, theories, and practice* (2nd ed.). Englewood Cliffs, NJ: Prentice-Hall.

Kagan, N. (1972). *Influencing human interaction.* Unpublished manuscript, Michigan State University.

Kanfer, F. H., & Gaelick, L. (1986). Self-management methods. In F. H. Kanfer & A. P. Goldstein (Eds.), *Helping people change* (3rd ed.) (pp. 283–345). New York: Pergamon Press.

L'Abate, L. (1981). Classification of counseling and therapy: Theorists, methods, processes, and goals: The E-R-A model. *Personnel and Guidance Journal, 59,* 263–265.

Linehan, E., & O'Toole, J. (1982). Effect of subliminal stimulation of symbolic fantasies on college student self-disclosure in group counseling. *Journal of Counseling Psychology, 29,* 151–157.

Norcross, J. C., & Prochaska, J. O. (1988). A study of eclectic (and integrative) views revisited. *Professional Psychology: Research and Practice, 19,* 170–174.

Patterson, C. H. (1985). *The therapeutic process: Foundations for an eclectic psychotherapy.* Pacific Grove, CA: Brooks/Cole.

Patterson, C. H. (1986). *Theories of counseling and psychotherapy* (4th ed.). New York: Harper & Row.

Prochaska, J. O., & Norcross, J. C. (1986). Exploring the paths towards integration: Ten ways not to get there. *International Journal of Eclectic Psychotherapy, 5,* 136–139.

Shertzer, B., & Stone, S. C. (1980). *Fundamentals of counseling* (3rd ed.). Boston: Houghton Mifflin.

Strupp, H. H. (1982). The outcome problem in psychotherapy: Contemporary perspectives. In J. H. Harvey & M. M. Parks (Eds.), *Psychotherapy research and behavior change* (Vol. 1, pp. 39–71). Washington, DC: American Psychological Association.

4

Cognitively Focused Counseling Strategies

INTRODUCTION

Four distinctive cognitively focused intervention strategies are presented in this chapter. These approaches can be seen as educative in nature since they help clients learn to do things in new or different ways. They have fairly clearly outlined techniques that, when used in appropriate situations, should bring about desired client goals. After studying this chapter you should be able to:

- discuss three ways that you can help clients acquire and retain factual information;
- describe how you might help a client learn the decision-making process;
- identify the characteristics that might indicate poor logical thinking on the part of a client;
- outline the steps that you might take to help a client restructure his or her thought processes; and
- name several ways that you could help a client think more analogously, inductively, or creatively.

Cognitively focused counseling strategies are based on the notion that individuals are rational beings whose cognitive and mediational processes influence all aspects of their ability to function. Consequently, counselors who want to help their clients change in significant ways need to use strategies that focus on the cognitive domain. This chapter discusses four major ways to work with clients from a cognitive point of view. These four approaches help clients (1) acquire and retain accurate and relevant factual information; (2) learn how to choose among various courses of action to make satisfactory decisions and solve problems; (3) learn how to think logically about themselves and the world around them; and (4) learn how to use their resources in creative, imaginative, and inventive ways. Each of these approaches contains different intervention strategies. This chapter outlines thoses strategies, suggests when these interventions ought to be used, and makes specific recommendations on how to employ these techniques.

ACQUIRING AND RETAINING FACTUAL KNOWLEDGE

Basic Concept

One major cognitive approach that is often employed, particularly by counselors who work in schools, colleges, and community agency settings, is helping clients acquire and retain factual information. Three specific strategies are used in this approach. They help clients

- learn more about educational and vocational opportunities,
- increase their knowledge about themselves by understanding the results of psychometric instruments,
- develop the skills necessary to manage the learning process more effectively.

Each of these strategies is discussed in this section. These strategies may be used alone, or they may be combined with other interventions to help individuals resolve issues and solve problems.

Educational and Vocational Information

One of the critical developmental issues that students face is the need for vocation selection and career development (Healy, 1982). Similarly, many adults need assistance in seeking employment opportunities (Herr & Cramer, 1988). Therefore, one important cognitive strategy that counselors are called upon to provide is assisting clients in obtaining relevant educational and occupational information. The choice of an occupation is one of the major decisions an individual makes in life. It affects the economic, psychological, and sociological activities of the individual. Economically, it determines certain financial rewards and, as Hoppock (1976) has pointed out, whether one will experience employment or unemployment, satisfaction or dissatisfaction, and success or failure. In the psychological and sociological domains, one's occupation provides a sense of identity and offers status and prestige; it strongly influences the selection of friends, leisure-time activities, who one marries, where one lives, and how one's children are educated.

Relevant educational and occupational information should be provided to students and clients to help them discover the variety of occupations available in our society; the similarities and differences among various occupations; what economic and psychological rewards are provided for by various occupations; which occupations are increasing in the number of those employed in it, and which ones offer decreasing opportunities for employment; what individuals actually do in the various occupations; the qualifications necessary for entry-level positions in these occupations; how one advances in these occupations; and how particular occupations affect other dimensions of life.

Fredrickson (1982), Isaacson (1985), Hoppock (1976), Tolbert (1980), and Zunker (1990) have all stressed the importance of counselors having an adequate and appropriate file of educational and occupational information. In general, they suggest that the counselor's file contain the following items:

> *Government publications* such as the *Occupational Outlook Handbook, Occupational Outlook Quarterly, Dictionary of Occupational Titles,* and the *Standard Occupational Classification Manual.*

> *An up-to-date file of pamphlets* on occupations, careers, and educational information. This material is probably best obtained by purchasing a complete set from one of the principal publishers of occupational information such as Careers, Chronicle Guidance Publications, and Science Research Associations. These publishers also provide a subscription service for keeping the file current.

> *Bulletins and directories* that are related to the educational and vocational interests of students and clients. These directories may include such items as bulletins for various colleges and educational programs, directories of various trade unions, directories of various employers, and the local telephone book.

> *Computer software* now contains considerable educational and occupational information. For example, the *Guidance Information System* (GIS) (1978) provides a procedure for obtaining national and local educational and occupational information through a time-sharing process. Other examples are the *DISCOVER* (1984) program and *SIGI PLUS* (1985), which provide systematic approaches for the career-exploration process. These programs are available in software packages for use with most of the popular desktop computers. Counselors who have computers and microcomputers should become familiar with the resources available in this medium. For a further discussion of this area, readers are referred to Fredrickson (1982), Isaacson (1985), Johnson (1983), Maze and Cummings (1982), and Zunker (1990).

The vast amount of educational and occupational information available will preclude you from having detailed knowledge about many occupations or educational programs. Consequently, you will rarely provide clients with the required information during an actual counseling interview. As a counselor who is concerned with the career development of your clients, you will need to become familiar with the available career-guidance resource materials and computer software packages. Furthermore, you should be able to instruct your clients in how to obtain this information so that it can be discussed in further sessions. The instructional process should include a discussion of appropriate resource materials, how the information can be obtained, how the material can be used by the client, and a timetable for accomplishing this search. Whenever possible, you should show the client how to use the resource library or the computer program or both.

The career development of individuals is facilitated by helping clients ascertain the implications of educational choices, the realities of the world of work, and the environmental factors that influence educational and vocational trends. Assisting clients in obtaining educational and vocational information includes accomplishing the activities specified in the following steps:

1. Orient the client to the availability of educational and vocational information. This should involve a discussion of:
   a. Educational and vocational opportunities that are available
   b. Sources of meaningful and reliable information
   c. Specific factors one should look for in seeking appropriate information
2. Help the client focus on several specific educational or vocational possibilities that he or she would like to investigate. This can be done in a variety of ways, including having the client:
   a. Provide a self-generated list
   b. Brainstorm with the counselor
   c. Take an interest inventory
3. Explain to the client that specific details are important and necessary to learn about each opportunity. For example, if the client is seeking information about a specific occupation, he or she should obtain information on the occupation's:
   a. Outlook
   b. Qualifications
   c. Rewards
   d. Work environment
   e. On-the-job activities
   f. Entrance requirements
   g. Advantages and disadvantages
4. Advise the client about the availability of resource materials on occupations and the world of work. This includes:
   a. Educational and occupational files
   b. Brochures, pamphlets, and books
   c. Computer programs
   d. Audio and visual resources
   e. Interviews with original sources
5. Help the client activate his or her search by providing:
   a. Instruction on how to use the resource materials
   b. Help in understanding the content of the information available
   c. Continued encouragement and positive feedback
6. Assist the client in examining this material in light of other needs. The client needs to assess this information in light of his or her:
   a. Background
   b. Aspirations
   c. Needs
   d. Decision-making skills

7. After the client has had the opportunity to evaluate this factual material, one of three outcomes is possible:
   a. The information is sufficient and the client needs no other help at this time
   b. Further information is necessary and the client needs continued help in obtaining it
   c. The information appears sufficient at this time, but the client needs help in decision-making skills or with some other problem

Counselors who work in school settings are called upon to be the experts in the school's developmental career-guidance program. Thus in addition to helping students on an individual or group basis, you may be expected to design and implement a career-guidance program that incorporates into the curriculum appropriate career information and decision-making skills. Counselors who work with clients on career decisions need to be thoroughly familiar with the contemporary theories of vocational development. An excellent summary of these theories is presented in Osipow (1983).

Interpretation of Psychometric Data

In addition to helping clients obtain good educational and vocational information, counselors often must assist clients who want to learn more about themselves and their characteristic traits. This process can be facilitated by employing appropriate psychometric instruments and interpreting these results in ways that are useful to clients. Before you discuss any psychological test result with a client, you should be thoroughly familiar with the instrument. You must know the validity and reliability of specific instruments; the types of scores that are available; and how to select, administer, and interpret appropriate tests. Important aspects of these concepts are presented in this section; however, a further discussion of these may be found in Anastasi (1988), Cronbach (1984), Goldman (1971), Mehrens and Lehmann (1984), and Walsh and Betz (1990).

The most important characteristic of a psychological test is its *validity*—that is, the degree to which it measures what it is intended to measure. A test may be valid for one situation when it is used for the purpose it was designed for and not valid in another situation when it is used for a purpose it was not designed for. Validity can and is expressed in several different ways: face validity, content validity, construct validity, and concurrent or predictive validity.

FACE VALIDITY. A term that is used when a test looks like it assesses what it claims to measure. Technically speaking, the term has nothing to do with whether or not the instrument does what it is purported to do. It is a useful notion, however, because people are more apt to take a test seriously if it appears to do what it is supposed to do. This type of validity is based on the opinion of the user and is purely subjective.

CONTENT VALIDITY.   A phrase that is employed when the instrument assesses a representative sample of the subject matter that it intends to measure. This type of validity is established by examining whether the items on the test truly represent an appropriate sample of the content areas. It is most often used when the domain or the universe of the subject matter is clearly defined. For example, standardized tests of academic achievement should have good content validity because academic subject areas are fairly well established. Test batteries that use this construct include the Stanford Achievement Test Battery and the Iowa Tests of Basic Skills.

CONSTRUCT VALIDITY.   The type of validity that is used when the instrument measures a theoretical psychological construct. This sort of validity is determined by examining the test to see if the items are logically consistent with the theory underlying the construct. The concepts of intelligence and personality traits are examples of psychological constructs. Examples of instruments that use construct validity include the Otis Lennon Mental Ability Test, which is based on a particular definition of intelligence, and the Edwards Personal Preference Schedule, which was designed to measure particular personality characteristics.

CONCURRENT VALIDITY AND PREDICTIVE VALIDITY.   Terms used when the instrument is compared to another independent criterion measure. The major distinction between the two terms is the time at which the criterion measures are taken. When the criterion measure is obtained at approximately the same time as the test data taken, concurrent validity is computed; when the criterion measure is obtained at a future time, predictive validity is acquired. Both concurrent and predictive validity are established by statistical means and expressed as a correlation coefficient. Values of this coefficient range from .00, or no relationship, to +1.00, or a perfect relationship. Acceptable instruments usually report correlation coefficients of validity of .60 or higher.

Concurrent validity is most important when the instrument is used as a more efficient way to obtain the desired information than a more involved method. For example, if a paper-and-pencil test of scholastic aptitude has a high correlation with an individual intelligence test, the first instrument can be used to obtain nearly the same information as the longer or second test. Thus, the Otis Lennon Mental Ability Test will yield scores that should be comparable to IQ scores obtained on the Wechsler Intelligence Scale for Children.

Predictive validity is most important when the instrument is used to help someone choose a future program or when a program wants to select the most promising candidates. Examples of tests that report predictive validity are the Law School Admission Test (LSAT) and the School and College Ability Tests (SCAT).

The Usefulness of Psychometric Data

A test is considered to be *reliable* when it measures the same phenomena in a consistent or stable manner. Because educational and psychological constructs are more dynamic than stable, and since any measurement of these constructs is a sample measure, reliability is usually thought of in probability terms. In other words, the reliability of a test can be thought of as the probability that the same person at *different times* or on a *different sample of equivalent items* will obtain the same score. Similar to concurrent and predictive validity, the reliability of a test is usually expressed statistically as a correlation coefficient. This statistic shows the degree of relationship between two sets of measures and varies from +1.00, a perfect relationship, to .00, no relationship. In general, a test should have a reliability coefficient of .80 or higher.

An individual who takes an educational or psychological test obtains a raw score on that instrument. This score, of and by itself, has no meaning unless the results of the test are associated with some standard. The standard may be a criterion reference or a population reference. A *criterion* reference is used when the score is based on a clearly defined content domain. Criterion-referenced tests are very useful for educational purposes in which mastery of certain knowledge and skills is deemed important. Note how important the criterion reference is in the following two examples: "Peter scored 85 out of 100 on an 8th grade vocabulary test"; "Susan typed 60 words per minute on a 6th grade vocabulary typing test." A *population* reference is employed when the score is compared to the scores of a representative sample of individuals called the *standardization sample* or the *norm group.* When the standard is a population reference, the raw score is converted to some relative score such as derived IQ, grade equivalent, percentile, stanine, *t* score, or *z* score. Each of these derived scores allows us to compare the individual tested to some reference group. For example, the statement "He scored in the 55th percentile on a 5th grade arithmetic test" really says, "He scored better than 54 percent of all fifth-grade students in the United States, and 44 percent scored better than he did."

Psychological tests provide a standardized and objective methodology for assessing certain educational and psychological characteristics of people. They can be used for classification, selection, evaluation, verification, and further understanding. They are used for *classification* purposes when the user wishes to classify individuals into some characteristic group that the test measures. For example, if a group of students was administered the Strong Vocational Interest Inventory, we could classify them by their highest scores according to the six personality types identified by Holland (1985)—namely, realistic, investigative, artistic, social, enterprising, and conventional.

Tests are used for *selection* purposes when the scores are used to select individuals for some purpose. For example, a company may wish to hire only individuals who have typing proficiency; thus, a minimum score on a standardized typing test for employment is set. Tests are used for *evaluation* when the user wishes to measure the results of an intervention, treatment, or teaching program. For example, a teacher wishing to evaluate the usefulness of a particular teaching modality may give a pretest and posttest to an experimental and a control group of students. Test are employed for *verification* when one wants to verify a scientific hypothesis. For example, a counselor wants to know if the self-concept of individuals can be improved by training the subjects in experiential focusing. To verify this hypothesis the counselor would have to set up an appropriate experimental design.

Finally, tests are used for *self-understanding* when a person learns more about himself or herself by taking a test. For example, a student may learn that her scores on the Strong Vocational Interest Inventory reveal interests very similar to accountants, an occupation she hadn't considered previously, and quite dissimilar to elementary-school teachers, an occupation she has seriously thought of entering. Although tests can be given for any of these five purposes, it should be noted that these purposes are not mutually exclusive and a particular test may be given for several reasons.

Several major points need to be made about administering educational and psychological tests in a counseling situation. First, careful thought must be given to selecting tests that measure characteristics that are important to the client. The instruments that are available can assess aptitudes, achievement, interests, or personality characteristics in different ways, and they all have distinctive purposes. Therefore, in selecting a test it is helpful to answer the question "How will the results of this test help the client?" Second, be sure the client is interested in taking the test and obtaining the results. Insofar as possible, involve the client in selecting the appropriate instrument. This requires careful explanation by the counselor of the strengths and attributes of any testing instrument under consideration. Third, explain who will have access to the results. Individuals have a right to privacy, and if any significant others, such as parents, have a right to see the results, that fact should be discussed with the client. Fourth, the directions for administering the test should be adhered to so that the test is always given with the same set of instructions. Counselors must be thoroughly familiar with the test manual and follow the directions specified. Respondents should know what is expected of them, how they should take the test, and how they should gauge their time. Finally, while the client is taking the test, any individual or environmental conditions that might affect the outcome of the test should be observed and recorded.

Relevant psychometric data must be carefully interpreted by counselors so that the information is useful to their clients. There are four basic types of test interpretation: descriptive, genetic, predictive, and evaluative. A *descriptive* interpretation helps the client learn what kind of person he or she is. For example, "Your scores on the Kuder Interest Inventory indicate that you prefer outdoor activities to computational ones." A *genetic* interpretation informs the client why he or she acts in a certain way. For example, "The discrepancy between your son's quantitative and verbal scores indicates a strong possibility that he has a learning disability, and that may be the reason for his poor school work." A *predictive* interpretation will help the client ascertain whether or not he or she is likely to succeed in a given endeavor. For example, "Students with scores like yours are usually accepted at a state college." Finally, an *evaluative* description helps the client compare himself or herself to some performance standard or criteria. For example, "Your score placed you in the seventy-fifth percentile. That means you did better than three-fourths of the students your age, and about one-fourth of the students did better than you."

A good testing procedure involves the client as much as possible. He or she should feel involved in the test selection and its evaluation and interpretation. The client can be involved in the interpretation process by following these four steps:

1. Encourage the client to share any relevant information about himself or herself with you. Try to ascertain: (a) the client's motivation for taking the test; (b) the conditions under which the test was taken; and (c) how serious the client was when he or she responded to specific test items and how the client felt about the process.

2. Tell the client about the nature of any of the test instruments that may be used. Describe what these tests purport to do, how the scores can vary from one administration to another, and how the results can be used. Whenever possible, involve the client in the selection of the testing instruments.

3. Interpret the results to the client using all the empirical evidence available. This may involve comparing the client's score to available statistical and normative information, and assisting the client in understanding his or her probability of success in pursuing a particular course of action.

4. Discuss the meaning of the test results with the client in light of other known factors. Use both test and nontest data to help the client learn more about himself or herself. Involve the client in the interpretation of the test. Have the client state what she or he believes the test score means in light of these other known factors. Allow the client to evaluate the test score as it applies to himself or herself.

The Management of Learning

In addition to helping clients learn about themselves from test data and obtaining educational and vocational information from good source materials, counselors often must assist clients in managing the learning process more effectively. Counseling programs at the primary and secondary education levels need to work with students from a developmental and preventive approach. Such an approach calls for helping students accomplish appropriate developmental tasks such as those specified by Havighurst (1972). In addition, counselors need to be attuned to the fact that several conditions influence the learning process (Griggs, 1983). These include:

- environmental conditions such as light, sound, and temperature;
- sociological characteristics such as peer and group pressure;
- perceptual strengths and weaknesses; and
- internal factors such as drive, persistence, motivation, and other psychological characteristics.

Helping students learn how to learn more effectively is an important component of a developmental counseling approach.

Furthermore, counselors who work in schools and colleges often must work in a remedial situation and deal with clients who manifest difficulties living up to their potential. This problem may be the result of a number of factors such as lack of interest, poor motivation, or overwhelming personal problems; or it may be the result of poor learning habits and skills. When the lack of academic achievement is related to the former concerns, the use of group or individual counseling procedures based on a combination of different intervention strategies is often called for (Baker, 1981). When the lack of achievement is related to learning deficiencies, counselors have to help clients manage the learning process more effectively.

In learning how to manage the learning process more effectively, students need to develop a variety of skills. Among these are learning how to:

- be prepared for their lessons with the necessary tools;
- allocate, schedule, and use their time effectively;
- concentrate and minimize external distractions;
- read textual materials;
- take notes from both oral and written presentations;
- organize their notes in systematic ways; and
- outline materials for oral and written presentations.

Green (1985) discusses a number of useful techniques for helping students improve their study skills.

Learning is notably affected by both external conditions and by a student's physiological and psychological status. External conditions such as the physical setting (light, sound, temperature, physical plant, and time of the day) and the sociological and cultural expectations of the groups the student belongs to (peer, family, and role models) create an environment that can enhance or detract from learning. An individual's physical health and perceptual strengths and weaknesses (auditory, visual, tactual, and kinesthetic) often influence how that individual progresses with certain educational tasks. And clearly a person's psychological characteristics (aptitudes, interests, motivation, and other personality traits) strongly shape how that person learns.

As a counselor you can help clients manage the learning process more effectively by using the following procedures:

1. Learn as much as you can about any conditions related to the client's physical, environmental, sociological, or psychological characteristics that may enhance or delimit the learning process. Investigate the possibility of modifying the environmental factors that hamper this process and note the student's strengths.
2. Encourage the client to develop habits conducive to learning. These minimally include:
   a. having the proper tools;
   b. working in an appropriate physical environment;
   c. learning how to use time wisely;
   d. discovering how to read and outline textual materials;
   e. seeking help and feedback as warranted; and
   f. using all available resources (for example, teachers and books).

3. Prepare the client for learning by facilitating the conditions that foster learning. These include:
   a. motivating the client by stressing how and why certain material is meaningful and important;
   b. providing feedback to help the client see logical relationships between the new materials and previously obtained information; and
   c. helping the client develop a useful system that can reinforce the learning process.
4. Help the client learn how to organize materials for learning more effectively. The client may have to be taught how to:
   a. group or categorize the information to be learned into associated clusters;
   b. arrange this material in a logical sequence that builds on previous knowledge;
   c. establish relationships between old and new materials; and
   d. use various mnemonic devices such as stories, sentences, words, jingles, and mental images to facilitate recall.
5. Assist the client in developing a plan for distributed practice. Help the client understand that three one-hour sessions are normally better than one three-hour session because the difficulty of learning increases disproportionally with time due to limits of attention span, fatigue, and environmental factors; and individuals tend to remember and recall the primary and most recent information.
6. Encourage the client to use a multisensory approach to learning. Have the client incorporate as many of the five senses as feasible, and teach the client to use a good study method such as the SQ3R method (Survey, Question, Read, Recite, Review).

In the process of learning, an individual establishes a pattern or set that facilitates recall and fosters learning similar concepts and facts. If the client has had a bad experience in learning or learned to do things incorrectly, the set must be modified through a reeducational process. This process is often harder to accomplish than the initial learning process because connections that have been made previously need to be changed so that a new pattern can be established and set. For example, a student who has learned to add a column of numbers incorrectly needs to unlearn this procedure as well as learn the correct way. As a result of continued or unwarranted failure or possible misuse of reinforcement procedures, an individual learns to become less competent, and a state of learned helplessness can develop. This set is a difficult but not impossible pattern to alter.

SYSTEMATIC PROBLEM-SOLVING AND DECISION-MAKING SKILLS

Basic Concept

Another major cognitive approach is one which teaches clients a decision-making process. The need for effective decision making is an ongoing process throughout the life span of an individual; it is also something everyone must do several times a day. Concerns may vary from simple daily tasks, such as "What shall I eat for breakfast," to more complex and difficult ones such as choosing a college, selecting an occupation, or picking a community in which to live. Choosing one thing over another can be confusing, stressful, and sometimes immobilizing. Counselors are often required to help their clients learn how to make decisions and solve problems. Clients need assistance in learning how to weigh facts, resolve conflicts, deal with frustrations, choose between alternative courses of action, cope with various life problems, and overcome indecisiveness.

Decision making is a procedure that can be learned and applied when one needs to solve a problem or make a decision. Typically, the decision-making process involves systematically working through a series of steps (Krumboltz & Thoresen, 1976; Tolbert, 1980; Yates, 1990); each step should be used in order to avoid premature action. These steps are:

1. orienting the person to the decision-making process;
2. identifying the decision that needs to be made;
3. generating a list of alternative courses of action;
4. gathering information relative to each alternative;
5. weighing the relative merits of each option;
6. making a choice among the alternatives;
7. implementing and evaluating the decision; and
8. generalizing the decision-making process.

The decision-making process involves a considerable amount of uncertainty and subjective self-evaluation on the part of clients. They should analyze their own needs and characteristics and relate these to the information they have collected or learned about each option. In assessing themselves clients often need to evaluate one or more of the following characteristics:

- the immediacy of their need to make a decision;
- their personal economic, social, and psychological needs;
- their personal beliefs and values;
- their previously learned pattern of making decisions;
- the degree of risk they are willing to take;
- their personal feelings about the seriousness and the importance of this decision;
- how this decision will affect other aspects of their lives; and
- how this decision will affect significant other persons in their lives.

This subjective evaluation of their previous adequacy in making decisions, how they have typically dealt with barriers and obstacles, and how their decisions have affected them and others, provides clients with an awareness of the dimensions of their choice process and the critical information necessary for making sound decisions.

Some clients have developed poor decision-making habits. Gelatt, Varenhorst, and Carey (1972) outline seven poor patterns or styles of dealing with the need to make a decision. In the first four patterns clients make decisions using inadequate procedures.

1. An *impulsive* decision occurs when clients use little thought or examination, often choosing the first available choice—"Leap before you look."
2. A *fatalistic* pattern is exhibited when the individual leaves the decision up to chance—"It's beyond my control."
3. A *compliant* procedure is manifest when the person follows someone else's plans—"I'll do whatever you say."
4. An *intuitive* decision is made when the client follows an inner feeling about the choice—"It feels right."

In the following three patterns clients are often unable to make a decision.

1. A *delaying* procedure occurs when the client postpones action until a later time—"I'll solve that one later."
2. An *agonizing* feeling happens when the person receives too much information or too many alternatives and is overwhelmed—"There are so many possibilities, I don't know what to do."
3. *Paralysis* occurs when the individual has the necessary information and accepts the need to make a decision but cannot face it emotionally because of tension or anxiety—"I can't face it."

The counseling process is a good vehicle for helping clients learn how to make appropriate decisions and learn the problem-solving process. Because of the similarity in the steps involved in reaching a decision and finding a resolution to a problem, the terms are often used interchangeably or the decision-making process is considered a form of the problem-solving process (D'Zurilla & Goldfried, 1971). The counselor's role in this process is fourfold. First, a warm, accepting atmosphere that is conducive for exploration and self-evaluation should be created. Second, the systematic steps necessary for making sound decisions must be taught or reviewed. Third, the client should be helped in learning more about the kind of information that he or she needs and the sources of that information. Fourth, the client must be assisted in considering each option in light of the information available and with due consideration to his or her own personal characteristics and subjective factors. It is, of course, the client's responsibility to reach a decision or solve a problem. This often needs to be made clear and reinforced with clients, particularly with those who like to procrastinate or those who want to foster a dependency relationship.

Individuals with low self-esteem often have difficulty with the decision-making process for several reasons. They may be unwilling or unable to take any sort of a risk, or they may feel that they have no control over their future and are unable to change things. They may be too sensitive to the feelings and ideas of others as well as to various social pressures and thus do what others believe is right. Such individuals are often poorly motivated and will question, either consciously or unconsciously, the need to study and learn the decision-making process (Brown & Lent, 1984; Janis & Mann, 1977; Wheeler & Janis, 1980). In such cases you will need to use other intervention strategies as well as teach the client the decision-making procedure.

Clients can and do differ in their personalities and in their decision-making styles. For example, some rely more on the objective data that they gather during the process; others rely more on their own subjective feelings and perceptions about the information. Some need considerable time to move through the process; others can move through it more rapidly. Some like to weigh their decision verbally, whereas others rely more on their own internal thought process. Clearly, you should be aware of these different styles and work with clients in their own modalities for the benefit of the client.

Teaching the decision-making procedure in the counseling process should lead to two outcomes. First, the client should be able to come to a resolution of the immediate problem that he or she is concerned about. Second, the client should learn the process so that he or she can generalize and apply the process to other situations that require decision-making skills. The process of using the decision-making procedure in counseling should be approached in a well-planned step-by-step approach. The following eight-step model illustrates such a procedure.

1. *Orientation.* Orient the client to the decision-making or problem-solving process. Help the client:
   a. recognize that decision-making and problem-solving skills are used daily and throughout life;
   b. understand that learning the process of decision making is necessary to solve immediate as well as distant problems;
   c. make a commitment for the necessary time and effort to learn these skills; and
   d. agree to act only after facts are weighed.
2. *Identify problem.* Help the client define the problem and his or her concerns in specific concrete terms. This step can be facilitated by:
   a. articulating the goals, values, and objectives as concretely as possible; and
   b. subdividing the overall objective into several subproblems or more manageable issues.
3. *Generate alternatives.* Encourage the client to generate a comprehensive list of possible alternative solutions. This process can be facilitated by:
   a. brainstorming;
   b. free-associating; and
   c. obtaining opinions of others.
   Initially the client should list as many alternatives as possible no matter how foolish some may first appear. This tends to ensure that no reasonable alternative will be overlooked. This list should then be arranged in order of preference.
4. *Collect information.* Instruct the client in ways of collecting information about each of the possible alternative solutions. The client may need help in developing appropriate questions and seeking the proper resources. Suggested resources include:

     a. interviewing knowledgeable persons;

     b. visiting appropriate sites;

     c. reading various resource materials; and

     d. engaging in exploratory activities.

5. *Evaluate alternatives.* Assist the client in examining the consequences of each possible or probable solution. The positive and negative consequences of each alternative should be examined in terms of the:

     a. rewards;

     b. obligations;

     c. probability of accomplishment;

     d. implications for self; and

     e. implications for significant others in the client's life.

6. *Plan action.* Help the client select and decide on a course of action. This often involves:

     a. eliminating the least desirable solution and selecting the most promising in terms of the consequences;

     b. reevaluating one's objectives in light of the implications of each solution;

     c. specifying the tactics or ways the alternative or strategy will be carried out (note what is to be done, when, by whom, how); and

     d. note the possibility of revising the decision if new information becomes available, or if an unexpected event occurs (new information and/or unexpected events are often cues to reconsider the entire decision-making process).

7. *Implement decision.* Encourage the client to act on his or her decision in order to evaluate and test the chosen alternative. If the solution proves to be satisfactory, then the decision process is complete; if not, then the process must be continued and the client must be assisted in going back to an earlier step.

8. *Generalize.* Assist the client in generalizing the problem-solving and decision-making model to other aspects of his or her life. For example, the model can be used:

     a. in everyday situations when one decides what to wear and how to spend one's leisure time;

     b. at critical developmental points in one's life when one decides what college to select or what occupation to pursue; and

     c. at stressful moments when one needs to decide on a course of action under painful or unpleasant circumstances.

Learning the decision-making model helps individuals anticipate problems, minimizes the probability of acting impulsively, and lessens the anxiety and tension often associated with crises and indecisiveness.

The Trait/Factor Approach

A unique form of the decision-making process is the trait/factor theory of counseling, which was developed to help clients make better educational and vocational choices (Williamson, 1965). This approach to counseling applies the decision-making process to help clients assess their *traits,* consider them in light of the *factors* in the environment, and reach some conclusion regarding their future behaviors. In this approach the counselor takes an active role in helping the client come to a better understanding of his or her abilities, interests, and values; the opportunities, requirements, and consequences of possible courses of action; and the steps necessary to accomplish an objective. The counselor assumes the responsibility for guiding the client through the following five stages of the counseling process: analysis, synthesis, diagnosis, treatment, and follow-up.

In the *analysis* stage, you should try to understand the client's concern or problem and learn as much about the client's traits and factors as possible. This information is obtained by means of psychological tests, interviews, and any other feasible means of collecting data. Next you need to *synthesize* this information into a meaningful pattern and make some *diagnosis* or prognosis regarding the meaning of these data for the client. In the *treatment* stage you should guide the client in his or her understanding and interpretation of these traits and factors, and suggest possible courses of action based on the available data. Furthermore, during the treatment stage the client is expected to come to some decision and plan to implement this decision in the foreseeable future. The *follow-up* stage takes place some time later when you check with the client to ascertain whether he or she is encountering any difficulty implementing the planned action.

The trait/factor approach to counseling appears to have been used mostly in helping clients make educational and vocational decisions (Patterson, 1986). Its use, however, is not restricted to these areas because it provides a systematic methodology for guiding counselors in dealing with a variety of issues (Ivey, Ivey, & Simek-Downing, 1987).

SYSTEMATIC IMPROVEMENT OF THE DEDUCTIVE THINKING PROCESS

Basic Concept

A third major cognitive approach deals with clients' deductive-reasoning powers, their perceptions of themselves, and their perceptions of the situations they are confronted with in life. This approach is based on the concept that the sources of clients' problems are the mistaken beliefs and perceptions they have about themselves and the negative constructs and connotations they have about their life experiences. The emphasis is placed on making clients aware of their mental processes; helping them eliminate unproductive and debilitating thought patterns, beliefs, and opinions; and enabling them to learn more adaptive, constructive, and useful patterns of thinking.

There are three important reasons why helping clients to think systematically and logically is important. First, expectations and assumptions have significant implications for emotional reactions and overt behavior in the real world. What you believe and what you expect to happen can influence how you feel and act. Second, a person's thought processes about an event or situation rather than the event itself are frequently the cause of erroneous feelings, distorted perceptions, or ill-advised behaviors. What you think about an event is often more important than the event itself. Third, because the process of thinking uses language as a reasoning tool, language itself plays an important role in a person's cognitive feelings and actions. Identifying certain things as "good," or "bad," or "so-so" will shape one's reactions to those items. Therefore, clients can be helped to modify their beliefs, attitudes, and behavior by changing the way they label things.

Clients' expectations or other assumptions about events, their logical thought processes, and how they label things are three important issues to address in helping clients overcome dysfunctional cognitions. Beck (1985), Ellis (1989), Mahoney (1974), and Meichenbaum (1977) offer distinctive approaches for helping clients who reach conclusions based on faulty reasoning processes. Often this reasoning contains one or more of the following distorted ways of speaking or thinking that reveals negative feelings about oneself.

Self-deprecating statements. These expressions constantly reveal poor self-worth. For example, "I'm not a good student" or "No one really likes me."

Absolute or perfectionistic terms. When an individual sets up overly stringent guidelines for his or her behavior, the individual sets himself or herself up for self-criticism and a negative self-image. Conclusions that are absolute or perfectionistic often include the words "must," "ought," "should," "unless," or "until." For example, "I should have been the one promoted" or "Unless I get an 'A' I can't go home."

Overgeneralization of negative experiences. These are deductions based on too few examples or situations. Frequently they are based on negative experiences that make clients think there are many obstacles making the future hopeless and bleak. For example, "Since I failed the first exam, I will fail the course" or "All the children in school hate me."

Negative exaggerations. These statements greatly magnify the true meaning of an event or reality. For example, "All professional athletes are greedy" or "You insulted my mother—you hate my family!"

Factually inaccurate statements. These remarks are based on inadequate or incorrect information. These erroneous data distort the client's perceptions of reality. For example, "You need an 'A' average to get into college" or "Autistic children are lazy."

Ignorance of the effects of time. These assertions ignore growth, maturation, and the effect that the passage of time can have on an experience or event. For example, "He was a very poor student last year—he will surely fail this year" or "I have to go back to the lake and relive my vacation there."

These examples reveal a faulty deductive-thinking process. This type of thinking can be analyzed by careful examination of the premises, facts, or assumptions involved and the conclusion reached from those facts or assumptions. The deductive-reasoning process involves at least two premises, factual statements, or assumptions to reach a conclusion that logically flows from the two premises. In analyzing a client's faulty reasoning, it can be very helpful to write out the premises, facts, or assumptions, and the conclusions based on these premises. Sometimes one or more, of the premises, facts, or assumptions is false. At other times the conclusion cannot be logically deduced from the premises. For example, using six of the statements from the previous section, the premises and conclusions could be reconstructed and written out in the following way:

1. My mother bawled me out.
   My father bawled me out.
   No one really likes me.
2. My parents expect me to do well.
   For me, to do well means to get an "A."
   I can't go home unless I get an "A."
3. Mary doesn't talk to me.
   John calls me the teacher's pet.
   All the children hate me.
4. Professional athletes make a lot of money.
   Those who make a lot of money are greedy.
   All professional athletes are greedy.
5. Autistic children have difficulty in learning how to read.
   Lazy children have difficulty reading.
   Autistic children are lazy.
6. He was a very poor student last year.
   Once a poor student, always a poor student.
   He will surely fail this year.

The first three examples are illustrations of faulty reasoning using premises or assumptions that may be true. The second example contains a statement that reveals a student's high academic expectation of self. The last three examples are prototypes of logical or syllogistic reasoning that is faulty due to erroneous assumptions. Generally, of course, the client's faulty logic is not stated in the syllogistic format but only as an abbreviated conclusion. Therefore, to help clients see the errors in their faulty thinking processes it is helpful to write down the premises and conclusions on a piece of paper or constantly point out the faulty premises leading to the distorted conclusions. Furthermore, clients sometimes need to investigate the evidence that supports their assumptions. This may require homework assignments. Two intervention strategies, thought stopping and cognitive restructuring, are very useful techniques when counselors want to help their clients' improve their deductive-reasoning processes.

Thought Interference or Thought Stopping

This technique is used to help clients recognize inappropriate thinking patterns and then terminate the unwanted thought (Cautela & Wisock, 1977; Mahoney & Arnkoff, 1978; Wolpe, 1958). It is often employed with clients who consistently present self-deprecating, negative self-concepts and poor images of themselves; exhibit unwarranted fear or anxiety in anticipating a future action; or display a strong emotional reaction to ordinary, everyday life situations. Clients may ruminate on the past, present, or future; engage in illogical or worry-oriented thinking; present self-distortions; or engage in self-destructive descriptions. This procedure is designed to restrain and inhibit the clients' negative self-talk and force them to speak in more realistic, accurate ways. Typical thought-stopping methodology involves the following steps:

1. Identify with the client one of the thoughts that the client would like to control, and explore the details surrounding the situations that seem to arouse the self-limiting thoughts.
2. Instruct the client to imagine himself or herself in a situation that raises irrational, poor, or self-defeating thoughts or thought sequences. If necessary, help the client verbalize the scene.
3. As soon as the irrational or poor thought is emitted by the client, interrupt with strong intervention—for example, "Stop! Your statement makes no sense."
4. Help the client change his or her inappropriate thought pattern and appreciate the reality of the situation by labeling the incident more accurately. The client needs to learn that any statement has four possible truth values: always, necessarily true; possibly true or true under certain circumstances; probably not true, true under rare circumstances; and never true, impossible.
5. To realistically effect changes in the client's cognitions, this process must be repeated with other visualizations until the client verbally changes labels and the direction of the thought patterns.
6. As with any other cognitive strategy, continuing guided practice should be combined with simulations until proper perspective is obtained. Homework should be assigned to obtain the experience and skills necessary for improving the thought process on one's own.

The thought-stopping procedure normally involves several sessions. In applying this strategy you should slowly shift the burden for interrupting the thought pattern from yourself to the client. This may be accomplished by gradually lowering your voice when you command the client to stop his or her thought; eventually ask the client to take the responsibility for saying "Stop!" whenever a disturbing thought occurs. Repeated practice is often necessary.

Systematic Cognitive Structuring and Restructuring

This technique is based on the underlying assumption that the way clients structure or think about their experiences determines how they feel and behave (Beck & Weishaar, 1989; Burns, 1980). Cognitive restructuring is designed to help clients modify maladaptive thought patterns and improve and extend their logical thinking process. It is often used when unwarranted conclusions are drawn because of erroneous premises or assumptions, or poor syllogistic or deductive reasoning, and when clients need help in extending their logical thinking process. Counselors who use the cognitive restructuring process employ the following procedures:

1. Take an extensive client history to find out how the client handles current problems and how she or he has handled them in the past.
2. Help the client become aware of his or her thought process. For each conclusion, ask the client to discuss the evidence for the particular conclusion, alternative ways of viewing the evidence, and what would happen if one or more of the alternatives were true.
3. Review the rational thinking process, showing the client the four possible truth values of any statement or assumption. Discuss and review how the thinking process works, using syllogisms when necessary. Show the client how the underlying assumptions, premises, and poor reasoning affect his or her thinking, feeling, and performance. It is helpful to present illustrations of irrational thinking in an exaggerated way so that clients can see them more easily.
4. Help the client analyze self-statements, assumptions about others, and logical thought patterns in light of the major dimensions of his or her life. A three-column procedure can help do this. In the first column the client should describe any and all anxiety-provoking situations. In the second column, the client can record his or her thoughts and conclusions about this situation. And in the third column the client should list the types of errors found in these thought patterns and conclusions.
5. Teach the client to modify internal self-statements, erroneous labels, and any poor assumptions about others.
6. Drill and review the logical-reasoning process with concrete cases from the client's frame of reference. When appropriate, help clients determine reasonably attainable goals. Having clients understand the importance of realistic goals decreases illogical thought processes.
7. Combine thought stopping with simulations, homework, and relaxation techniques until logical patterns become set.

Occasionally clients face difficulties extending their thinking processes. When this occurs, clients need help understanding the interrelationships between and among complex constructs and data; the relationship of past, present, and future events; or how environmental factors are related to current behavior or events. This often requires factual-instruction techniques rather than cognitive restructuring.

The thought-stopping and rational-restructuring techniques may be useful with individuals who tend to be extremists (that is, perfectionists, overly dependent, isolationists, or irresponsible types); with those who exhibit fears and anxiety in regard to performance in social situations, public speaking, or test taking; and with those who display extreme emotional reactions to normal life situations (that is, depression or anger).

ANALOGOUS, INDUCTIVE, AND CREATIVE THINKING

Basic Concept

A fourth major cognitive approach is one that encourages clients to be open to new thoughts and ideas and to think in creative ways (Cole & Sarnoff, 1980). Individuals are frequently called upon to act or make a decision when it is not possible to use the deductive-reasoning process to reach positive certain conclusions. They must learn to think analogously, inductively, and creatively as well as deductively. They connect one case to another and learn from this comparison. They generalize from individual cases and learn to act in creative ways. This type of thinking provides clients with conclusions that go beyond observations or known facts. It is helpful to foster analogous, inductive, and creative thinking when we want to motivate clients to try to do something new; raise their aspirational levels; gain new perspectives on common everyday problems; and obtain some awareness or insight into or about themselves.

Individuals reason analogously when they compare two or more things, find a resemblance in some respects between and among them, and therefore conclude that the items are probably alike in other respects as well. *Analogous thinking,* in other words, is a process of reasoning by comparison. Individuals find a similarity among objects and experiences and reach a conclusion regarding the relationships between or among other objects and experiences. For example, a counselor-in-training has helped an undecided college freshman make a choice of college major by using some occupational information, having the client interview several individuals, and helping the client sort out the meaningfulness of these data. The next time the novice counselor has a client with a similar problem, the counselor will compare his or her previous experience to this new experience to see if there is enough in common to use the same intervention strategy.

Individuals think *inductively* when they generalize from a limited set of observations. A person starts with a number of observations, attempts to integrate or find a conceptual way to connect these observations, and then comes to some generalized conclusions based on both the observations and some thoughts about the relationships among these observations. For example, a new counselor has encouraged ten of his clients to discuss their personal interests with him. Subsequently, he administered an interest inventory to each of them. From these ten cases he has concluded that although discussion of the students' expressed interests provided useful insights into their backgrounds, the measured interests provided additional information relevant to a discussion of potential occupations. Thus, the counselor decided that he should use both expressed and measured interests when he does career counseling.

Creative thinking is very often employed when a solution is sought for a problem that cannot be resolved by the deductive-reasoning process alone. To free the creative potential that is inherent within each person, various methods have been suggested for generating these new thoughts. These include the use of metaphors (Gordon, 1978), lateral thinking (DeBono & DeSaint-Arnaud, 1983), and imagery (Wolpin, Shorr, & Krueger, 1986). Typically the creative process involves recalling one's previous experience with similar problems; using both analogous and inductive thinking to brainstorm and obtain new visual images of situations, problems, or issues; and breaking out of the usual or traditional ways of doing things in order to reach a new and perhaps unconventional solution to the problem. Creative thinking requires a fairly thorough knowledge of the field one is thinking about. For example, a counselor who has studied cognitive, affective, and behavioral approaches to counseling and who has experience using a variety of techniques during his or her training soon discovers that each client is different and that although one can employ a particular intervention, the application of that intervention often requires novel approaches and some creative thought.

Although using analogous, inductive, or creative reasoning with clients in certain types of counseling situations may be beneficial, these approaches have not received much attention in the counseling literature (May, 1984). It is believed that this kind of reasoning progresses through a definite four-stage process that provides the client with a useful understanding of this type of thinking.

The four stages in analogous, inductive, and creative thinking are: preparation/observation, incubation/analysis, illumination/tentative conclusion, and verification. During the *preparation/observation* stage, the problem or concern is identified, the facts and observations are gathered, the material is studied in detail, and questions are posed regarding the observations made. During the *incubation/analysis* stage, time is allowed for pondering all the observations and facts previously made, the data are compared and contrasted, and ideas and concepts are thought of and reviewed until the components of a problem are reformulated into a solvable issue. During the *illumination/tentative conclusion* stage, provisional solutions are reached. This stage often requires considerable time to reach and frequently occurs when one is not consciously forcing it. The *verification* stage occurs when one tries out a tentative solution, revises it if need be, and/or generalizes to other examples (Bernard & Huckins, 1975; Moore, 1985).

Suggested Methods

Counselors who want to foster analogous, inductive, and creative thinking in their clients can use a variety of methods. You should help the client move through the observation, analysis, conclusion, and verification process. The following four questions may be used by clients as guides to facilitate this process:

- What do I want to observe in this process?
- What meaning does this information have for me, my family, and my friends?
- What inferences, suggestions, or conclusions can I tentatively draw from this process?
- How can I try out these notions to see how realistic they are?

The seven methods suggested below can be used by counselors to encourage clients to be open to new ideas, to gain new perspectives, and to think in different ways.

1. Encourage creative problem solving. Have clients use the brainstorming technique to generate as many ideas as possible. Explain this concept and then ask some open-ended questions to foster imagination and new lines of thought. Allow the ideas to be a little wild, fragmented, unconnected, and unconventional. Instruct clients to defer judgment on all of the notions. The emphasis is to break out of the typical ways of thinking and open new thought patterns.

2. Try having clients dream, think, or write about ways to solve a problem or about the future. Try to have them look at questions addressing the future, such as: "In two years what would you like to be doing?"; "In three years how would you look back on this problem?"; "In four years what issues will you face?"; or "In five years where would you like to live?"

3. Use the open-ended-question technique to help clients think more openly. Provide the client with a printed list of unfinished sentences regarding his or her life. Examples might include the following: "I wish that . . ."; "If I could I would . . ."; or "In what ways could I . . . ."

4. Help clients understand that many activities and the conclusions we are called upon to make are not based on deductive thinking. Individuals often resolve problems by using analysis or by viewing the problem from another perspective. Ask your client how he or she might solve a problem that was somewhat different, or ask what suggestions the client would make if the problem belonged to a friend.

5. Encourage clients' use of imagery, relaxation, and meditational processes. Have clients learn the relaxation procedures and then the meditational and imagery techniques to become more aware of themselves (Cole & Sarnoff, 1980).

6. Encourage clients to interview possible role models. Help them identify possible interviewees who might provide them with some future direction or help them resolve an issue. Assist clients in formulating appropriate questions and, if possible, role-play the interviews. After the interviews are concluded, help clients integrate this experience and reach new conclusions.

7. Instruct clients to use various references to investigate alternative solutions to a problem. Suggest books and reading materials, audiovisual materials, or various computer software packages. Help them understand that there are often alternative ways to resolve an issue. Assist them in the process of integrating this material into a meaningful conclusion for themselves.

Summary

This chapter has outlined four cognitively focused intervention strategies. These strategies are based on the concept that human beings are rational and their cognitive and mediational processes exert a strong influence on all their endeavors. Cognitively focused strategies are designed to help clients acquire and retain accurate and relevant factual information; learn how to choose among alternative courses of action to make satisfactory decisions and solve problems; learn how to think more logically about themselves and the world about them; and learn how to use their resources in more analogous, inductive, or creative ways. In these strategies the counseling process is primarily an educative process, and as the counselor your major function will be to help the client learn new material and different ways of handling various situations. These intervention strategies typically involve a series of structured steps and frequently require clients to do a considerable amount of work outside the counseling session.

After reading this chapter on cognitively focused strategies you should be able to describe the major cognitive interventions. You should have an understanding of when these strategies can be used and recognize the steps involved in employing these methods. To attain proficiency in applying these methods in counseling sessions will require more in-depth study of these techniques and considerable clinical practice with actual clients under the supervision of a qualified professional.

REFERENCES

Anastasi, A. (1988). *Psychological testing* (6th ed.). New York: Macmillan.

Baker, S. (1981). *School counselor's handbook: A guide for professional growth and development.* Boston: Allyn & Bacon.

Beck, A. T. (1985). *Cognitive therapy.* In H. J. Kaplan & B. J. Sadock (Eds.), *Comprehensive textbook of psychiatry* (pp. 1432–1438). Baltimore: Williams & Wilkins.

Beck, A. T. & Weishaar, M. E. (1989). Cognitive therapy. In R. J. Corsini & D. Wedding (Eds.), *Current psychotherapies* (4th ed.) (pp. 285–320). Itasca, IL: F. E. Peacock.

Bernard, H. W., & Huckins, W. C. (1975). *Dynamics of personal adjustment* (2nd ed.). Boston: Holbrook.

Brown, S. D., & Lent, R. W. (1984). *Handbook of counseling psychology.* New York: Wiley.

Burns, D. (1980). *Feeling good: The new mood therapy.* New York: Morrow.

Cautela, J. R., & Wisock, P. A. (1977). The thought-stopping procedure: Description, application and learning theory interpretations. *The Psychological Record, 2,* 255–264.

Cole, H., & Sarnoff, D. (1980). Creativity and counseling. *Personnel and Guidance Journal, 59,* 140–146.

Cronbach, L. J. (1984). *Essentials of psychological testing* (4th ed.). New York: Harper & Row.

DeBono, E., & DeSaint-Arnaud, M. (1983). *The learning-to-think coursebook.* Larchmont, NY: DeBono Resource Center.

DISCOVER: A computer based career development and counselor support system. [Computer program]. (1984). Iowa City, IA: American College Testing Foundation.

D'Zurilla, T. J., & Goldfried, M. R. (1971). Problem solving and behavior modification. *Journal of Abnormal Psychology, 78,* 107-126.

Ellis, A. (1989). Rational and emotive therapy. In R. J. Corsini & D. Wedding (Eds.), *Current psychotherapies* (4th ed.) (pp. 197–238). Itasca, IL: F. E. Peacock.

Fredrickson, R. H. (1982). *Career information.* Englewood Cliffs, NJ: Prentice-Hall.

Gelatt, H. B., Varenhorst, B., & Carey, R. (1972). *Deciding: A leaders guide.* Princeton, NJ: College Entrance Examination Board.

Goldman, L. (1971). *Using tests in counseling* (2nd ed.). New York: Appleton-Century-Crofts.

Gordon, D. (1978). *Therapeutic metaphors.* Cupertino, CA: Meta Publishers.

Green, G. W. (1985). *Getting straight A's.* Secaucus, NJ: Lyle Stuart.

Griggs, S. A. (1983). Counseling high school students for their individual learning styles. *The Clearing House, 56,* 293–296.

Guidance Information System Guide. [Computer program]. (1978). West Hartford, CT: Time Share Corporation.

Havighurst, R. J. (1972). *Developmental tasks and education* (3rd ed.). New York: David McKay.

Healy, C. C. (1982). *Career development: Counseling through the life stages.* Boston: Allyn & Bacon.

Herr, E. L., & Cramer, S. H. (1988). *Career guidance through the life span* (3rd ed.). Boston: Little, Brown.

Holland, J. L. (1985). *Making vocational choices: A theory of careers* (2nd ed.). Englewood Cliffs, NJ: Prentice-Hall.

Hoppock, R. (1976). *Occupational information* (4th ed.). New York: McGraw-Hill.

Isaacson, L. E. (1985). *Basics of career counseling.* Boston: Allyn & Bacon.

Ivey, A. E., Ivey, M. B., & Simek-Downing, L. (1987). *Counseling and psychotherapy* (2nd ed.). Englewood Cliffs, NJ: Prentice-Hall.

Janis, I. L., & Mann, L. (1977). *Decision-making: A psychological analysis of conflict, choice, and commitment.* New York: Free Press.

Johnson, C. (Ed.). (1983). *Microcomputers and the school counselor.* Alexandria, VA: American School Counselor Association.

Krumboltz, J. D., & Thoresen, C. E. (1976). *Counseling methods.* New York: Holt, Rinehart & Winston.

Mahoney, M. J. (1974). *Cognition and behavior modification.* Cambridge, MA: Ballinger.

Mahoney, M. J., & Arnkoff, D. (1978). Cognitive and self-control therapies. In S. L. Garfield & A. E. Bergin (Eds.), *Handbook of psychotherapy and behavioral change* (2nd ed.) (Part IV, pp. 689–722). New York: Wiley.

May, R. (1984). *The courage to create.* New York: Bantam Books.

Maze, M., & Cummings, R. (1982). *How to select a computer assisted guidance program.* Madison, WI: University of Wisconsin, Vocational Studies Center.

Mehrens, W. A., & Lehmann, I. J. (1984). *Measurement and evaluation in education and psychology* (3rd ed.). New York: Holt, Rinehart & Winston.

Meichenbaum, D. H. (1977). *Cognitive behavior modification: An integrative approach.* New York: Plenum.

Moore, L. P. (1985). *You're smarter than you think.* New York: Holt, Rinehart & Winston.

Osipow, S. H. (1983). *Theories of career development* (3rd ed.). Englewood Cliffs, NJ: Prentice-Hall.

Patterson, C. H. (1986). *Theories of counseling and psychotherapy* (4th ed.). New York: Harper & Row.

SIGI PLUS: A computerized guidance program. [computer program]. (1985). Princeton, NJ: Educational Testing Service.

Tolbert, E. L. (1980). *Counseling for career development* (2nd ed.). Boston: Houghton Mifflin.

United States Department of Commerce, Office of Federal Statistical Policy and Standards (1980). *Standard occupational classification manual.* Washington, DC: United States Government Printing Office.

United States Department of Labor. (1977). *Dictionary of occupational titles* (4th ed.). Washington, DC: United States Government Printing Office.

United States Department of Labor. (1990). *Occupational outlook handbook: 1990–1991.* Washington, DC: United States Government Printing Office.

Walsh, W. B., & Betz, N. E. (1990). *Tests and assessments* (2nd ed.). Englewood Cliffs, NJ: Prentice-Hall.

Wheeler, D. D., & Janis, I. L. (1980). *A practical guide for making decisions.* New York: Free Press.

Williamson, E. G. (1965). *Vocational counseling.* New York: McGraw-Hill.

Wolpe, J. (1958). *Psychotherapy by reciprocal inhibition.* Stanford, CA: Stanford University Press.

Wolpin, M., Shorr, J., & Krueger, L. (Eds.). (1986). *Imagery.* New York: Plenum.

Yates, J. F. (1990). *Judgement and decision making.* Englewood Cliffs, NJ: Prentice-Hall.

Zunker, V. G. (1990). *Career counseling: Applied concepts of life planning* (3rd ed.). Pacific Grove, CA: Brooks/Cole.

PUBLISHERS OF CAREER INFORMATION

Careers, Inc., Largo, FL 33540

Chronicle Guidance Publications, Moravia, NY 13118

Science Research Associates, Inc., 155 Wacker Drive, Chicago, IL 60606

5

Affectively Focused Counseling Strategies

INTRODUCTION

This chapter is devoted to a discussion of affectively oriented counseling approaches. The person-centered, the Gestalt, and the existential approaches are briefly outlined. Suggestions are offered for using the affective approaches to help clients ventilate, obtain catharsis, and gain insight and improve self-awareness. After careful review of the material in this chapter you should be able to:

- discuss the key concepts in the three affective approaches: person-centered, Gestalt, and existential;
- identify the differences among the ventilation, the cathartic, and the self-awareness and self-insight processes; and
- describe how you could use an affective approach to assist a client in gaining insight into himself or herself.

Individuals sometimes want or need to explore feelings about themselves, their self-perceptions and self-responsibilities, their relationships with significant others, and the meaningfulness of their experiences. The concerns they present may reveal a positive desire to learn more about themselves, or they may present issues that reveal limited awareness of self, negative feelings of self-worth, feelings of helplessness or of having few options, or an inability to deal with their emotions in an effective manner. When any of these are present, clients may be helped by using an affectively focused counseling strategy.

Affectively oriented approaches are based on a view of human nature that emphasizes the following major points. First, human beings have the capacity to understand themselves, to regulate their own behavior, to choose among alternative courses of action, and to develop their inherent potential. Second, individuals view the world from their own subjective frame of reference. Third, healthy individuals live in the present and deal with phenomena at the conscious level. Fourth, the ultimate aim of any counseling process is to help clients move in a positive direction to find self-satisfaction through living a meaningful and authentic life. And fifth, any and all attempts to facilitate client growth must be based on a deep respect for the client as a human being, an understanding of the importance of the client's unique perception of the world, and a focus on or concern with conscious phenomena or the "here-and-now" events of life.

This chapter presents three major affectively oriented counseling theories. Each of these approaches helps clients gain a better understanding of themselves and thus become better able to deal with their problems, issues, and concerns. The *person-centered* theory developed by Carl Rogers (1951) emphasizes helping clients focus inwardly to help them move toward the process of self-development and self-actualization. The *Gestalt* theory founded by Fritz Perls (1988) focuses on assisting clients in improving their self-awareness by becoming more cognizant of the totality and the meaningfulness of their experiences. Individuals who become more aware of their experiences learn to function more effectively. The *existential* approach has been promulgated by several different theorists including Frankl (1985), May (1986), and Van Kaam (1966). This school emphasizes helping clients focus on the meaningfulness of their experiences to obtain a better understanding of their existence. Through this process clients gain more positive feelings of self, and more self-actualization or insight about self is likely to occur. After the presentation of the theories, the text outlines specific suggestions that can be employed to help clients ventilate, obtain catharsis, and develop greater insight and self-awareness.

Because each of these three major theories is reviewed only briefly in this chapter, readers are encouraged to read about them in detail either in the books written by the original theorists (Rogers, 1972; Perls, 1988; May, 1986) or in texts that present another description of the essential concepts of these theories (Burks & Stefflre, 1979; Corey, 1991; Corsini & Wedding, 1989; Gilliland, James, & Bowman, 1989; Gladding, 1988; Ivey, Ivey, & Simek-Downing, 1987; Patterson, 1986; Shilling, 1984).

THE PERSON-CENTERED APPROACH

An important affectively oriented counseling approach is the person- or client-centered one. This humanistic approach emphasizes that every human being has within herself or himself the potential to be creative and behave in a responsible and wholesome manner. The client is the central focus of his or her own therapy, and it is the client who ultimately reaches his or her own conclusions about what is causing any disturbance or problem, the steps that need to be taken to resolve the issue, and the consequences for living with any therapeutic outcome.

The person-centered approach to counseling was developed by Carl Rogers. It is based on the underlying premise that human beings, by their very nature, are inherently good; that they have a tendency to act positively; and that they can and do strive to become self-actualized. The emphasis in person-centered counseling is placed on developing an atmosphere that is conducive to client growth rather than on diagnosis or prescriptive interventions. In his writings

(1942, 1951, 1957, 1972) Rogers has stressed the following major concepts, which outline the frame of reference he used to understand how people function and how to assist them. These concepts are the dignity and worth of each person; the inherent potential of each person to become self-actualized; the importance of each individual's private world of experience; the natural characteristic of people to be good and trustworthy; his notions of self-concept and the feelings of positive regard that must be present for individuals to function effectively and fully; and the role and function of the counselor.

THE DIGNITY AND WORTH OF INDIVIDUALS. Rogers maintains that every person is a valuable member of the human race and deserves to be treated in a dignified and humanistic way. Each person, regardless of looks, economic status, or behavior, should be respected and treated warmly and genuinely. Having this nonjudgmental, positive, and genuine attitude toward others is essential for dealing with others in all relationships.

SELF-ACTUALIZATION POTENTIAL. Rogers stresses that every person, by his or her very nature, has the inherent potential to become self-actualized. This self-actualization process is the prime reason clients change and grow in positive directions. It implies that individuals have the innate capacity to function in ways that satisfy their physical and psychological needs in an autonomous but self-controlled and self-disciplined manner. This process serves to enable the person to develop those capacities that will enhance the individual and significant others in his or her life. Self-actualization is a dynamic process of becoming (Rogers, 1972), and it is a growth and developmental phenomenon rather than a static terminal state or fixed stage of existence. Self-actualized individuals are dynamic, function fully, and live effectively in socially responsible ways.

INTERNAL FRAME OF REFERENCE. A third major concept articulated by Rogers is the importance of understanding the individual's internal frame of reference. Each person lives in an ever-changing private world or field of experience that they alone know and encounter. This subjective or phenomenological world provides a frame of reference for the individual that in turn acts as a guide for the behavior of the individual in satisfying his or her perceived needs. Thus, an understanding of this perceptual world is basic to understanding a client's behavior.

TRUSTWORTHINESS. Rogers believes that individuals are intrinsically good and trustworthy. By this concept he means that all human beings have an innate tendency to manage their own affairs and behave in ways that are personally satisfying and socially responsible. Individuals who behave in irresponsible, socially undesirable, or destructive ways have become alienated from their own inherent nature. Their behavior is a defensive process that projects a poor self-image or poor self-concept. This underlying discrepancy within the individual causes anxiety, and the individual is in a state of incongruence.

SELF-CONCEPT AND POSITIVE REGARD. Rogers postulates that two inter-related factors influence the self-actualization process. The first is the person's self-concept. This self-concept is one's awareness of one's being. This view of self is composed of the constellation of one's attitudes, feelings, and thoughts about one's self. It includes concepts of one's abilities, skills, personality, goals, desires, values, relationships with others, and experiences. The self-concept is subjective and personal.

The second factor is the degree of positive regard that one has had and continues to experience. All human beings have the need to be accepted, re-spected, and loved, but it is of critical importance during the childhood devel-opmental years. Positive regard is a valuating concept; the degree to which one experiences this positive regard influences one's self-concept and thus the way one functions and one's tendency toward self-actualization. If during the de-velopmental period a child receives affection based only on his or her behavior, then the child experiences *conditional positive regard,* which can and does cause problems within the person. For positive mental health it is essential for the child to experience *unconditional positive regard,* a basic ingredient in Rogerian counseling.

Maladaption comes about as a result of experiencing conditional positive regard and a blockage of this natural tendency toward self-actualization. Rogers has labeled this as an incongruity that has developed within the person. This *incongruity* is often reflected in one of the following ways: a discrepancy be-tween the individual's awareness and his or her functioning; a discrepancy between the individual's potential and his or her attainment; negative feelings about one's self or one's characteristics; strong feelings of guilt, tension, or anx-iety; or poor interpersonal relationships.

THE ROLE AND FUNCTION OF THE COUNSELOR. Rogers has stressed that because clients have this inherent potential to self-actualize, the counselor's role is to facilitate this process by providing the appropriate climate so that self-exploration, self-understanding, and personal growth can occur. In this process the self-actualizing tendency of the client will help the client overcome any blockages or obstacles that have interfered with the person's ability to function. Rogers (1957) has stated that certain necessary and sufficient conditions must be present for positive client growth to occur. Five of these conditions focus on the counselor and one focuses on the client. These conditions are:

1. *The individuals must be in psychological contact.* This first condition places the responsibility on the counselor for creating a nonjudgmental, non-diagnostic, and nonauthoritarian atmosphere in which a relationship based on mutual respect and trust can develop. In this warm, genuine, and authentic climate the counselor should encourage the client to explore his or her own feelings, attitudes, and prescriptions; help the client focus on the present rather than past occurrences; and accept whatever the client says without a value judgment.

2. *The client is in a state of incongruence, being vulnerable or anxious.* The second condition indicates that the client is truly a person in need and is quite anxious, disturbed, or upset about some events, persons, or things in his or her life. This incongruence can manifest itself in a variety of ways. In some cases clients may distort or deny their feelings, perceptions, or some factual events, in other cases they may not be able to deal with experiences in their lives as they believe they ought to be able to deal with them. This anxiety may appear as a relatively minor concern or an extremely serious one.

3. *The counselor is congruent or integrated in the relationship.* The third condition requires that the counselor be receptive to the client and focus intensely on what the client is saying. The counselor must leave any personal concerns out of the relationship and not allow any barriers to interfere with listening to the client's words and being attuned to the client's underlying messages. This congruence allows the counselor to pay close attention to the client and to be genuine in his or her contact. Congruence further allows the counselor to be aware of self in the relationship and communicate in an open, sincere, and meaningful way. This in turn facilitates the client's willingness to open up to the counselor.

4. *The counselor experiences unconditional positive regard for the client.* The counselor needs to feel a deep and genuine caring for the client as a person of value no matter what condition the client is in, how the client behaves, or what the client says. The counselor does not have to like the condition, the behavior, or the expressions but must accept the client as a fellow human being who is entitled to respect without any conditions. This unconditional positive regard is a nonpassive caring for the client, and it encourages a mutual respect that is essential for the collaborative nature of person-centered counseling.

5. *The counselor experiences an empathic understanding of the client's internal frame of reference and endeavors to communicate this experience to the client.* Through the process of listening to the client's words and the underlying messages, the counselor senses what the client is feeling and what meanings these feelings have for the client. The counselor shares his or her own perceptions of these feelings with the client in an open, respectful, and sharing manner. This sensing of the client's inner world may be limited at times, but the communication of this intention has considerable value. It can foster further discussion of the client's hidden self and encourage more awareness of self.

6. *Communication to the client of the counselor's empathic understanding and unconditional positive regard is achieved to a minimal degree.* The client hears that the counselor really cares, understands his or her problems and concerns, and is aware of the client's underlying feelings and perceptions. This results in the client feeling a sense of genuine interest and concern; security, which allows a gradual discussion of inner feelings, attitudes, and perceptions; and being a mutual partner in the counseling process. The focus of the person-centered approach is not to help a client solve a particular problem but rather to help the person grow and develop a better understanding of self. Through the process of providing acceptance, respect, and understanding, the client

slowly allows his or her inherent self-actualization process to be activated. Clients also gain increasing insight, become more aware of the meaning of their own thoughts, feelings, and behavior, and begin to see for themselves how they can deal more effectively with the problems and issues of their own lives.

The counseling process goes through the following phases:

1. the experience of being cared for and the sense of freedom to discuss anything that is on one's mind;
2. a slow unfolding and airing of one's attitudes and perceptions, and a release of one's pent-up feelings;
3. a gradual movement to become less defensive or anxious;
4. awareness of one's incongruities and the factors related to these issues;
5. a more accurate perception of one's self, one's problems, and one's relationships to others;
6. gaining the strength to deal with these problems and an increased ability to make decisions; and
7. a gradual but definite sense of being more integrated, an increased facility for self-direction, more self-confidence, and a greater positive insight into self.

The process of uncovering feelings that have been distorted or denied in one's awareness is not always a smooth path but rather a route that has numerous ups and downs. In gaining insight and gradual self-awareness, the client may have to recognize some internalized patterns about his or her concept of self. Thus, the process for some clients is painful and anxiety provoking. Genuineness, unconditional positive regard, and an empathetic understanding of the client are essential conditions necessary to guide the client through this process.

THE GESTALT APPROACH

A second major affectively oriented counseling method is the Gestalt approach. The Gestalt approach was developed by Frederick S. Perls (1988; Perls, Hefferline, & Goodman, 1977). This counseling approach is process oriented and focuses on what is happening to the client during the counseling rather than on the content of the client's discussion. Client feelings are emphasized, and the therapeutic goal is to help clients become more aware of what they are doing and how they are doing it, and at the same time, to learn to accept and respect themselves and their own behavior. Perls used concepts from Gestalt psychology to gain a better understanding of the spectrum of human emotions, feelings, and body sensations. Gestalt counseling emphasizes the following major concepts: the wholeness of individuals; the ability of individuals to focus on their perceptual fields and to select one need at a time in a spectrum of needs, then shift focus from one need to another as needs are met; the inherent tendency of individuals to self-actualize and to direct their lives in an effective manner; the importance of the present; the homeostatic or self-regulating process; the concept of awareness; and the intervention techniques used in the counseling process.

WHOLENESS. The essential ideas of Gestalt psychology are based on the notion that human behavior is more than a mere collection of unrelated stimuli or events and that the parts of life have meaning in relation to the whole. Thus, individuals do not see phenomena as isolated bits and pieces. Rather, we have a tendency to organize these factors into meaningful configurations. The human experience cannot be compartmentalized; it needs to be understood as a *gestalt,* integrated, or unified whole (Van de Reit, Korb, & Gorrell, 1980). The original Gestalt theory was formulated in Germany by Wertheimer, Kohler, and Koffka, who were studying the field of perception (Hartmann, 1974). The Gestalt psychologists found that human beings have a strong need to impose a meaning on, and achieve closure with, relevant portions of their perceptual fields. Three major principles guide this need: (1) the principle of closure, which says that people have a need to complete the incomplete; (2) the principle of proximity, which says that the distance between objects has a bearing on how they are seen; thus, people tend to relate items that are close to one another and not relate items that are farther from one another; and (3) the principle of similarity, which says that people have a tendency to group similar items together. Visually these three principles can be represented by the following three examples:

- What shape is suggested by this diagram? | ‾‾‾‾‾‾ |

- How do you organize the following sets of dots?   • • •   • •   • •

- What figures would you group together?   □ ○   □ ○

Each one of these principles shows that individuals constantly seek to put things together in order to make sense, or to develop an organized pattern out of a variety of factors, and that individuals need to find a connection or to have closure in dealing with life's issues.

The concept of wholeness also refers to the agreement between the physical and the psychological dimensions of a person. When clients reveal discrepancies between verbal statements and nonverbal behavior, a lack of wholeness is present. Gestalt counselors look for evidence of this lack of wholeness and make their clients aware of these discrepancies, thus challenging their clients to assume responsibility for this incongruence.

CONCEPTS OF PERCEPTUAL FIELD. Another important notion is that individuals organize their perceptions to meet their needs. The perceptual field of an individual is described by Gestalt psychologists in terms of the *figure* and the *ground.* Within the individual's environment or perceptual field, the item that the person is focusing on or is most aware of is called the figure, and the rest of the perceptual field is called the ground. Only one item or event can be the figure at one time. The object that the client focuses on can be the figure, and the ground is anything within the client's knowledge and awareness levels. A good gestalt, or meaningful perception pattern, is created when a person's greatest need becomes the figure and other needs fade into the background or the ground. When this need is met, the gestalt is completed, and that figure fades into the ground allowing the individual to go on to another need. If needs are not met, gestalts are incomplete, new gestalts cannot be made, and the client has some *unfinished business.* Furthermore, when gestalts are not complete, or closure has not occurred in previous experience, the individual's self-regulatory process has malfunctioned. Thus, a hidden agenda for many individuals is their unfinished business. This unfinished business, which is often an unexpressed feeling or unresolved situation, will interfere with the effective functioning of the individual.

SELF-ACTUALIZATION. Perls (1988) stressed that human beings are inherently neither good nor bad, but they can use their nature effectively or ineffectively. All have the potential to self-actualize. This self-actualization process is manifested by the ability of individuals to find their own way in life and to accept responsibility for directing their own lives in an effective manner, coping with the problems and frustrations of life, and choosing between and among alternative courses of action. Therefore, a self-actualized person takes responsibility for his or her actions, lives a productive and proactive life rather than a reactive or defensive one, and thoughtfully selects from among options to meet his or her unique needs in a systematic manner. The ineffective or maladaptive person has not taken responsibility for his or her decisions and has not acted on appropriate figures in his or her perceptual field.

THE IMPORTANCE OF THE PRESENT. An important tenet of the Gestalt approach is concentration on the here and now. People must live in the present for it is the only time one has control over. The focal point of life is the here and now; the past is gone and the future has not yet come (Polster & Polster, 1973). Consequently, individuals need to appreciate and fully experience themselves as they live life and not dwell on the past or the future. The past and future are of importance only if they are related in a meaningful way to the present. In the counseling process this concept is stressed by encouraging clients to relate their feelings to the present and by not allowing clients to ruminate or focus on past or future events. Focusing on the present facilitates the client's awareness of self and enhances introspection, exploration, and resolution. This here-and-now orientation is fostered by the counselor who uses questions such as "What are you feeling right now?" and "What are your feet doing while you are talking?"

SELF-REGULATION. All human beings are constantly striving to meet their needs and are in an interactive relationship with their own unique environments (Latner, 1984). This relationship is a homeostatic process, or a tendency toward balancing, wherein the individual strives for equilibrium within himself or herself or between the self and the environment. This balance can be threatened by demands of the environment (outside of self) or by demands of various needs of the individual (internal conflicts).

When an imbalance occurs, often a strong emotional experience, such as anxiety or exuberance, arises and indicates that the individual is out of balance.

This emotional state usually signals the need for self-regulation and self-control, and normally supplies the energy for the organism to meet this need and return to a balanced state. When this fails to happen, the regulatory process has malfunctioned. This imbalance may be related to different spheres of life or to different intrapsychological phenomena. Examples of the former type of imbalance include how a student's poor interpersonal peer relationships influence her academic performance and how a person's home life affects his job performance. Examples of the latter type of imbalance include how the need for recognition becomes a self-defeating behavior in one's attempt to relate to significant others.

Gestalt theory outlines four major mechanisms whereby individuals become imbalanced: introjection, projection, confluence, and retroflection. *Introjection* refers to ways clients handle the "shoulds," the rules and regulations of life. Individuals who introject uncritically adopt behaviors that others say they should accept. This causes individuals to function poorly because they automatically accede to societal demands and act and behave as others expect them to behave or act. They are not their own persons. *Projection* is the process of disowning parts of one's self that are inconsistent with one's self-image and placing these parts in the outside world. One makes the environment responsible for what occurs within oneself and one's life; thus, one is not accountable for one's behavior. *Confluence* occurs when the client cannot make a clear distinction or boundary between himself or herself and the environment. A parent who considers a child an extension of self has invested so much of self in the child that the parent loses a sense of where he or she stops and where the child begins. *Retroflection* occurs when a behavior that is directed toward others is not successful and the person then directs the behavior inward, or when one cannot obtain what one wants from another so the person gives it to himself or herself. Thus aggression toward another person can be turned inward and the individual winds up hating himself or herself.

AWARENESS. The objective of the Gestalt counselor is to help the client become fully aware of self and his or her present experience. Self-awareness is of major importance in an individual's self-regulation process. It is perhaps the single most important feature of the self-actualized person. This is the central concept of the Gestalt approach. Awareness touches virtually every aspect of life: one's internal ideas, feelings, and physiological actions, as well as one's appreciation for and relationship to one's environment. When one is aware of self, one is in touch with one's own existence and reality; one has a sense of responsibility for one's own attitudes, behaviors, and feelings; and one can exercise some control over himself or herself. Helping the person become aware of self at every instant assists the client in seeing his or her difficulties, what may be producing them, and what actions may be necessary to solve them.

The aware person understands things as unified wholes or gestalts. When awareness is blocked, needs can be unmet, feelings unexpressed, self-actualization frustrated, and gestalts incomplete. These incomplete gestalts cause considerable imbalance within the individual and serve as a continual distraction for the person. People become maladjusted because they are fragmented and unable to determine which objects or phenomena in their environments will satisfy their needs. The major sign of a maladjusted person is being unaware of self or environmental factors or both. Energy is spent suppressing unfinished business. Hence, the key to self-actualization is self-awareness, and the major way to help individuals overcome their maladaptive behaviors is to increase their self-awareness.

THE ROLE AND FUNCTION OF THE COUNSELOR.   The aim of the Gestalt counselor is to assist clients in becoming more aware of themselves, their environments, and their personal needs in the here-and-now situation. Thus, the counselor helps clients come to grips with their emotions, feelings, and attitudes, and discover what is causing an imbalance in their lives. This increased self-awareness in turn makes the client more self-reliant and hence more self-actualized. In other words, the aim is to foster self-dependency by helping individuals learn about their own strengths and self-regulation activities. Gestalt therapy is not interested in why a person behaves the way he or she does but on what a person is doing and how he or she is doing it.

The Gestalt approach attempts to go beyond clients' verbal statements by helping them experience themselves in the present and work through their unfinished business. Client self-awareness is enhanced by encouraging clients to listen to what they themselves are saying; recognize the processes of what they are feeling and thinking; observe how they act in certain situations; and note any discrepancies between their thoughts, feelings, expressions, and behaviors. Rather than providing answers to the client's concerns, the Gestalt counselor encourages the client to deal with the unfinished business by structuring the situation so the impasse comes into the open. The counselor's goal is to help the client recognize the unfinished business or the impasse, realize that he or she has the ability to resolve the issue, and assume the responsibility of taking the necessary steps to act on the unfinished business. The counselor encourages clients to make their own interpretations and find their own meanings as they struggle to experience the process of completing their gestalts. By experiencing these problems and solutions, clients are able to increase this awareness and improve their self-regulation processes.

The role and function of the counselor in this process is to provide an atmosphere that is conducive to developing self-awareness in the here-and-now situation (Passons, 1975). Any techniques can be used to facilitate this process. Typically, Gestalt counselors are rather active, directive, and forceful in trying to help the client focus his or her awareness on the present. The task of the counselor is to bring out this awareness by attending to the client and his or her perceptual field, heightening or highlighting the essential elements of these experiences, and when necessary, expressing feelings that are too abstract or personal for the client to verbalize. It is not the role of the counselor to pass judgment on the client but rather to point out discrepancies, to challenge incomplete gestalts, and to give feedback.

Although Gestalt counseling does not mandate a specific step-by-step procedure, this approach does stress having clients experience themselves in problem situations. Gestalt counselors emphasize having clients take ownership of their own statements, act out conflicts, and see different sides of the issues. Although specific techniques are not mandated, Gestalt counselors frequently use the following strategies.

1. *Help clients modify their verbal statements.* In order to stress to clients that they must live in the present, take ownership over their lives, and take responsibility for their feelings, behaviors, or attitudes, counselors direct their clients to modify their verbal messages using one of the following techniques:

   a. Use the personal pronoun. The client is encouraged to use "I" and to avoid the impersonal "it." Speaking in the first-person singular forces the person to take ownership of his or her statements and enhances the client's awareness of what is being communicated. For example, a client is instructed to change "It will not happen again" to "I will not let that happen again."

   b. State that they are responsible for themselves. One way to do this is to have clients end all their statements or beliefs with the phrase "and I take responsibility for my feelings, actions, or thoughts." For example, a client is told to change "I was mad" to "I was mad, and I take responsibility for my feelings." Another technique is changing the words "can't" to "won't" or switching "but" to "and." For example, "I want to stop smoking, but I can't" is a statement that avoids responsibility. "I want to stop smoking and I won't" puts the responsibility for the action in the client's lap.

   c. Talk in the present tense. This encourages the client to focus on the here-and-now situation. For example, clients are asked to change "I hated him" to "I hated him yesterday, and today I am still very angry with him."

   d. Convert questions into statements. Forcing clients to make statements encourages them to focus on themselves and their own perceptions and beliefs and avoids deflecting the focal point away from the client. For example, "Do you believe me?" should be changed to "I do not think you believe me."

2. *Use simulations.* Simulated techniques are used to help clients gain insight into their own behaviors, emotions, perceptions, and thoughts. By using these techniques they can learn to bring past experiences into the present, understand both sides of a question or situation, feel the opposite sides of an internal conflict, or get in touch with a part of themselves they may have been unaware of. These simulations are also used when the client has some unfinished business that is limiting the client. Such techniques as role-playing, role reversal, rehearsing, dialoguing, playing the projection, and the empty-chair technique can be employed. After acting out any of these simulations, the counselor helps the client see the unfinished business that emerges.

a. Role-playing occurs when the counselor has the client play himself or herself in a given situation or relationship. It provides the client with the opportunity to engage in a behavior without any risks. Gestalt counselors often ask clients to say out loud or enact what they are thinking about.

b. Role reversal happens when the client plays someone else in a given situation or relationship. It gives the client the chance to see himself or herself and the situation from another perspective. One technique used is asking the client to play part of himself or herself that he or she rarely or never expresses.

c. Rehearsal is the act of practicing a specific behavior or situation. It affords the client the opportunity to strengthen the behavior and belief so that the target behavior can be carried out.

d. Dialoguing is the process of speaking on both sides of an issue or a conflict. This brings the issue out into the open and provides the client with new insight.

e. Playing the projection occurs when the client has projected something on another person and the counselor has the client play the role of the person on whom the projection was made. Thus clients may see that they have the same questions as the other person.

f. The empty-chair technique is similar to the dialogue, but it is more dramatic because the process involves having the client talk to an empty chair, that stands for a significant other person or the opposite side of an issue. This technique is sometimes carried out by having the client change seats each time he or she speaks on a different side of the issue.

3. *Direct clients to discuss unpleasant feelings and emotions.* Gestalt counselors do not preach or use "why" questions, but they do direct clients to focus on unpleasant experiences. This is done to help the client experience the emotion during the counseling process. The threatening fantasies are brought out into the open, and the client is encouraged to see how he or she can handle the obstacles. Strong resistance to this is often encountered, and the counselor needs to point out how the client is stuck at a prohibitive point. This awareness of being "stuck" helps the client overcome his or her resistance to the process. One of the following methods is often used:

a. Asking clients to finish open-ended sentences such as "Right now I feel . . ." or "I would like to . . ."

b. Encouraging clients to stay with a feeling and even exaggerate their emotional experiences during the counseling session—for example, "Stay angry, get real mad at him, show me how angry you can be" or "Cry, go ahead and cry if you feel that sad." One variation of this is to ask the client to become the emotion: "Be your anger, stay with it."

    c. Confronting clients with discrepancies between their statements and nonverbal gestures or body language—for example, "You say you are relaxed, but you are biting your nails."

    d. Encouraging clients to share how they might feel if they had just revealed a well-guarded secret to a group or to a particular person— for example, "Suppose you just told your innermost feelings to the people you work with. How would you feel?"

    e. Sharing your hunches about how you think a client might feel about a given situation—for example, "It seems to me that you would feel very hesitant to tell your boss off."

THE EXISTENTIAL VIEWPOINT

The third major affectively oriented counseling approach is based on the existential notion that human existence, by its very nature, causes individuals to search for the meaning of their lives. Thus, many of the problems that clients bring to the counseling process are related to their search for identity and a purpose in life, their examination of their relationships with others, and their questioning of their own abilities and developmental processes. The existential approach does not offer specific techniques for working with clients; rather it recommends issues that counselors ought to be aware of and endorses the creation of an atmosphere conducive for client growth in examining the meaningfulness of life and its ramifications.

    The existential approach to counseling does not represent the work of a single person, nor is there one major proponent of this school. It has its roots in existential philosophy, which was originally discussed and written about by such men as Kierkegaard, Nietzsche, Sartre, and Heidegger. A number of individuals such as Frankl (1985), Fromm (1989), May (1989), Van Kaam (1966), and Yalom (1981) have been influential in advancing the existential approach in counseling. These experts have articulated a number of characteristics that distinguish those who are living a meaningful existence from those who are functioning inadequately. Among the major characteristics of those who live an authentic existence are awareness of self; a sense of freedom to make responsible decisions; the ability to relate to significant others in their lives; the willingness to accept the negatives of life and death; and the acceptance of anxiety as a constructive force in one's life.

    SELF-AWARENESS AND UNIQUENESS. Individuals have the innate tendency to be aware of themselves, a unique capacity that allows them to think and feel about themselves. This awareness is primarily a self-centered subjective perception or the "I-am" experience (May & Yalom, 1989). This awareness facilitates an understanding of the fact that one is unique in the world, and the process of learning underscores the experience that each person travels the road of life in a way that no one else does. This quality of aloneness can lead to loneliness and isolation. When clients feel this strong sense of isolation, a prime therapeutic goal is to help them become more aware of themselves and their uniqueness. Increased awareness of self leads to greater courage to face life's loneliness, greater responsibility, greater openness to options, and richer existential living. Failure to be aware of self leads to passivity and reactive lifestyles rather than productive and proactive ones.

FREEDOM AND RESPONSIBLE DECISION MAKING.   Individuals have the capacity to make choices in their lives, and they are responsible for their actions. Human beings are essentially free to make decisions and to take alternative courses of action in many dimensions of their lives. These choices are often made with a degree of risk and uncertainty. Existentialists hold that individuals grow, learn, and develop in a process that does not simply follow a behavioral-conditioning process. People are responsible for the choices they make and the ones they refuse to make; thus, they are responsible for their behavior and the direction of their lives. Individuals who do not use their capacity to make decisions, or who are fearful of the consequences and thus become immobilized, act as if their lives were controlled completely by external forces. When clients are immobilized by conflicts, the major existential goal is to help them understand that freedom is a basic human quality.

RELATIONSHIP.   Individuals have a basic need to relate to significant others in a meaningful way. The need to be a part of a group, the need for recognition from others, and the need to share oneself with others are all manifestations of this need for relatedness. Individuals who are open to others show a caring and respectful attitude toward them and feel responsible for their well-being. Individuals who do not have a good relationship with others become estranged and alienated from society. This lack of relationship can also lead to isolation and self-centeredness. The therapeutic process aims at helping the client become more aware of the need to relate and the interdependence of all members of the human society.

MEANINGFULNESS.   Human beings have an innate desire to search for a personal meaning for their existence. This desire to understand the meaningfulness of existence is the major characteristic of existentially oriented approaches. Individuals who are functioning effectively accept the mysteries of existence and the dualities of that existence—birth and death, joy and sadness, elation and pain, and happiness and suffering. This acceptance does not imply a passivity nor any avoidance of the desire to ameliorate or improve the negative aspects of that existence but rather that life, by its very nature, has these components to it. Having a sense of the meaning of life gives a purpose to one's existence and enables one to form a set of moral and spiritual values. Searching for this meaning in the face of the sufferings and dilemmas of life is an ongoing existential quest. The existential counseling process helps clients deal with these issues and learn to resolve these ultimate questions about life.

THE ROLE OF ANXIETY. Anxiety is natural to the human condition. Rather than viewing anxiety as a negative state of imbalance, individuals must learn to use anxiety as a constructive awareness to act. Living life fully can and does cause anxiety. Choices must be made with elements of risk. Responsible actions can cause unpleasant reactions on the part of others, and intensive and worthwhile task accomplishment normally has concurrent pressures. Anxiety is not something to avoid; rather, it should motivate the individual to take proactive, positive, and responsible steps to deal with the causes of the anxiety. Individuals who avoid anxiety have relinquished control of themselves and lead a reactive and passive existence.

THE ROLE OF THE COUNSELOR. Existentially oriented counselors focus on helping clients recognize that seeking meaningful life satisfaction from material objects and physical things is a nonauthentic lifestyle and that true satisfactions come from living life in an existential manner (Fromm, 1955). Thus, existentialists tend to help their clients become more

- aware of their actions and their ability to choose alternatives and be responsible for these actions;
- aware of themselves, their uniqueness, loneliness, and independence;
- sensitive to the factors that influence their relationships to others;
- open to their quest for searching out a meaningful purpose or value in life; and
- accepting of anxiety as a healthy human experience that naturally occurs in living one's life.

Ultimately, the goal of the existential counselor is to help the client experience a more authentic life.

The existential counselor must see the client as a unique person who needs to map out a role for self in this world. Thus, the concern is with helping the person become free to follow his or her authentic self and face the challenges of life. Existentialists do not employ a step-by-step procedure to accomplish these objectives, nor do they offer a set of practical techniques. Emphasis is placed on developing a sound working relationship and creating an honest, open, and nonauthoritarian atmosphere. The counselor is actively engaged in giving caring messages: expressing authentic responses, helping the client gain insight into his or her own thoughts and feelings, and aiding the client in dealing with the values and meaning of life.

Existentialists can and do use a wide variety of methods to accomplish these objectives. Any methodology that helps the client focus on understanding a personal meaning of life, that has a distinct emphasis on personal responsibility, and that helps the client explore his or her own notions of self and the process of living may be used. The crucial concept is developing a warm, trusting relationship with the client. This relationship process is the key to enabling the client to become more authentic in his or her own life.

AFFECTIVELY ORIENTED COUNSELING PROCEDURES

Affectively oriented counseling strategies direct the counseling process inward, that is, toward the client in order to help the client deal with his or her own perceptions and feelings about self. This inward attention is based on the notion that clients have the inherent ability to function more effectively, and that the counselor's role is to act as a catalyst to allow and encourage this process. Although various experts use different processes to assist their clients, the following major counseling strategies generally employ affective approaches. These intervention strategies foster ventilation, catharsis, and insight and self-awareness, and thus help clients gain improved feelings and perceptions of themselves. In general, the affectively focused intervention strategies include the following concepts or guidelines.

- Deal with the client in an open, respectful, and genuine way.
- Focus on understanding the client from his or her subjective internalized frame of reference.
- Work with the client on a conscious, focused, here-and-now orientation.
- Accept the client as having tremendous dignity and worth, regardless of his or her present condition.
- Focus on developing client self-awareness by using techniques that encourage the client to see himself or herself more clearly, use reflective and self-awareness procedures to enhance this process.
- Help the client understand self, his or her underlying motivation, and the environmental and personal factors that influence his or her behavior.
- Assist the client in seeing alternative ways of thinking, feeling, or behaving; this will tend to enhance the client's sense of freedom and responsibility.
- Help the client understand how he or she can influence the direction and activities of life and enhance self-responsibility.
- Help the client explore personal values and the meaningfulness of life experiences, search for greater self-fulfillment, and become more self-actualized.

Ventilation

BASIC CONCEPT. One important affectively oriented counseling strategy is ventilation. *Ventilation* can be defined as the process of allowing clients to talk about things that are of concern to them. This process of providing clients with an outlet to examine, discuss, and investigate their feelings, thoughts, opinions, and experiences is an important counseling strategy and will often make clients psychologically healthier (May, 1989). Creating a warm, accepting climate allows clients the freedom to discuss whatever is of concern to them in a nonthreatening way and allows them to release inhibitions, makes possible a flow of internal feelings, and helps them see things more objectively. Although fostering this atmosphere is an essential ingredient for all counseling processes, it is often appropriate to focus on this skill or climate on its own.

Focusing on the ventilation process is often very useful when clients are under some stress due to the following conditions: indecisiveness, a recent trauma or negative experience, or a recent profound positive experience. Ventilation can be useful to clients who know how to make decisions but who need help to reflect on various components of themselves; their relationships with family, friends, and significant others; and the decisions they need to make in the foreseeable future. This process can also be of considerable value to clients whose functioning level is normally quite satisfactory but who need to talk about a recent profound or negative experience. The ventilation allows clients to express themselves openly and air their feelings, and thereby find release from the tension they are experiencing. Thus they become more free to make decisions and to regain their normal level of functioning.

The ventilation process is facilitated by providing the fundamental counseling conditions articulated by Rogers (1957). Create an empathic, warm, and respectful climate conducive to client openness and self-disclosure. Allow the client the opportunity to experience and explore the full range of whatever is bothering him or her. Try to feel and understand the client from his or her internal frame of reference by intensively listening with the "third ear" and reflecting on the meaning of the client's statements rather than the surface content. Respect the client as a human being of significant worth and value. And encourage the client to talk about things that are difficult for him or her.

Attend to the client. Actively listen to what the client has to say. Manifest your active listening by using a variety of attending skills such as accents, simple minimal verbal responses, paraphrases, reflections of content and feelings, and summarizations. Attend to the client's internal frame of reference rather than to surface statements. Concentrate on listening to what the client is saying and the meaning of these messages, and communicate back to the client that you know and understand what he or she is trying to communicate.

Clarify when necessary. When some confusion or doubt exists about what the client means or is trying to communicate, or when the client presents double messages, ask an appropriate question that is designed to clear up the confusion, resolve the doubt, or identify the duality of an issue more clearly. Clarifying is often needed to define what the real issue is. Perception checks such as "It seems to me that you are trying to say this . . ." and questions clarifying alternatives such as "Do you mean this or do you mean that?" are helpful communication tools that can facilitate the clarifying process.

Support the client as a person of significant value. Provide the client with positive feedback about himself or herself. Inform the client that he or she seems like a fine person to you. Communicate feelings of reassurance and encouragement, and thus reaffirm the client's sense of personal value. This support or reassurance is frequently communicated by expressing approval of a present, past, or future action or feeling; or by providing consolation for some unhappy event. Helping clients relax by using a warm, friendly tone or by employing specific relaxation techniques is another supportive process that can facilitate the ventilation process.

Use open-ended questions or statements when appropriate. Active listening is also facilitated by the use of open-ended statements and questions such as "Tell me more about that" or "What happened then?" The judicial use of these open-ended statements and questions, posed with a warm, understanding tone, encourages further self-exploration and discussion.

Catharsis

BASIC CONCEPT. A second affectively oriented counseling strategy is catharsis. Catharsis is thought of as a more complex and focused intervention than ventilation. It is a purging or cleansing action that results from a release in tension or the elimination of an emotional blockage by bringing it out into the open. This process is based on the concept that one's emotions can control the way one thinks and the way one acts, so that feelings that haven't been expressed build up inside until the pressure or tension causes one to behave in maladaptive ways (Bohart, 1980). These pent-up feelings keep one from functioning adequately, use up a considerable amount of internal energy, and often result in avoidance of situations, withdrawal from life, and keeping oneself at a considerable emotional distance from significant persons in one's life. These unexpressed feelings may be relatively recent but often are based on traumatic experiences that occurred long ago. The buried feelings associated with that experience need to be expressed so that the built-up tension can be discharged and the pressure within the person restored to normal.

In cases in which emotions are related to deep feelings of bitterness, anger, and frustration, discharges can be dramatic and possibly explosive. Catharsis can be used to help relieve grief, fear, embarrassment, rage, anger, boredom, and tension (Scheff, 1979). The cathartic process can be facilitated, and it can occur by encouraging the ventilation process. Catharsis may be enhanced further by helping clients remember and express the circumstances under which their emotions became blocked, and by observing and understanding the discrepancies in their lives that may have caused the emotional difficulty.

The cathartic process unfolds over two phases. First is the recall of the past experience, and second is the release of the blocked emotion in one form or another (Nichols & Efran, 1985). The first stage is facilitated by encouraging the ventilation process and allowing clients to recollect their experiences. The second phase not only allows the pent-up emotional release but also focuses on helping clients deal with these emotions. During this release and recovery phase clients should be assisted in the following ways:

- encourage them to relive their suppressed traumatic emotional experiences;
- foster the examination of these experiences so that they understand why the blockage occurred and what ramifications this suppression has caused for self and for others;
- help them understand how this experience can be seen as a valuable part of their lives; and
- encourage them to explore how they can overcome any problems or interpersonal-relationship difficulties that may be related to this blockage.

No attempt should be made to foster other counseling strategies at this time. Allow the client to discuss whatever he or she thinks is of importance. Do not respond with solutions to the content of the client's problem, but recognize the issues as real concerns. Do not expect to hear complete stories or logical sequences. The ventilation and cathartic processes are often erratic, emotionally charged, and disjointed. During the ventilation process do not feel that you, as the counselor, need to know the whole story or foster a resolution to the issues presented. During the cathartic process, the release of tension and pent-up feelings should be the primary emphasis. The object of these strategies is to allow the client to ventilate and to experience the release of the tension or stress blocking his or her ability to function. Sometimes the ventilation process alone can improve the client's functioning levels. At other times the cathartic process seems necessary. These strategies can take place over one session or unfold gradually over several interviews.

Insight, Self-Awareness, and Improved Self-Esteem

BASIC CONCEPT. A third important affectively oriented counseling strategy focuses on helping the client gain insight and improved self-awareness. Insight is the process of looking inside oneself to gain a deep understanding of one's own existence. People can gain insight into one of three major dimensions of their lives. First, they can gain a more complete knowledge and awareness of themselves. Second, they can obtain a better understanding of the relationship between themselves and significant other persons in their lives. And third, people can explore and come to a better understanding or acceptance of the meaningfulness of their experiences in the world. Helping clients acquire insight in any or all of these three dimensions facilitates self-awareness.

In the process of understanding themselves, clients need to gain a more accurate analysis of their own personality characteristics and psychological attributes. They may have a confused, distorted knowledge of themselves and need help to overcome this confusion and distortion. They may need assistance in learning to become more aware of themselves and need help to:

- discern what thoughts and beliefs are held superficially and which ones are held deeply;

- distinguish between or among various feelings and attitudes;
- detect their needs, interests, and motives with a deeper understanding; and
- obtain a view of their strengths and weaknesses from another perspective.

Obtaining insight also has a dimension of becoming more aware of one's relationship to significant others and how one's own thoughts, feelings, and activities affect these relationships. Individuals can be aware of the images they present to others and the reasons why people relate to them the way they do; or individuals may present themselves in a distorted manner in the belief that the distortion is necessary to maintain the relationship. The counseling process can foster insight and self-awareness by helping individuals:

- understand how their self-perceptions affect their relationships with others;
- learn how they feel about others and how they believe they treat them;
- discover the importance of their relationships with others; and
- perceive how these behaviors affect their social existence, personal growth, and satisfaction.

Fostering insights has existential connotations as well as personal and interpersonal ones. The individual who is insightful and has self-awareness should be aware of the realities of life and feel responsible for self, others, and the well-being of society. Being aware of the meaningfulness of one's experience implies an openness to the existential world and an acceptance of the vicissitudes and temporalness of life. Counselors often help clients gain insights into the meaningfulness of their experiences by assisting them in

- understanding their own values and beliefs about the world;
- becoming more accepting of unpleasant, painful experiences;
- realizing the limitations of the human condition; and
- striving to become open in their experiences and the realities of their existence.

Clients who can benefit from an insightful-oriented approach typically manifest a lack of self-knowledge, a subjective view of themselves that is incongruent and confused, or a self-image that is not in agreement with the views of other significant persons. They may present themselves as shy, docile, or confused people whose behaviors or conversation reveals confused, unproductive, counterproductive, or self-defeating ways. The picture clients portray of themselves may be accurately described by them, or they may be only vaguely aware or unaware of this lack of self-knowledge. Other clients may present themselves as individuals who function at a fairly high level but need help sorting out the meaning of their existence, or they may be suffering from a traumatic experience and may benefit from the insightful counseling process.

Insight and self-awareness is typically the second phase of an intervention strategy that employs the ventilation process as the first stage. In other words, providing the facilitative conditions, or a warm, open, and respectful climate, is necessary to enhance insight. Often, the skills that are used to foster the ventilation process are sufficient of and by themselves to allow insight to occur. Rogers (1972) has presented both theoretical and research findings to support his claim that his facilitative conditions are necessary and sufficient for positive changes to occur. However, although the Rogerian conditions may be all that are necessary for some clients under certain circumstances, additional techniques are available and can be used to foster insight.

The insightful process can be developed by a variety of techniques that enable clients to see reality more accurately, trust self-decisions more certainly, and understand others more completely. One or more of the following communication skills may facilitate this process.

Provide the client with appropriate factual information or explanations of how various structures work or how they may be organized. Frequently clients lack knowledge about themselves, environmental factors, and available resource material. If some important knowledge is lacking, insight may be facilitated by providing this essential material. The information may be presented by the counselor, or the counselor may help the client find it by directing the client to it and guiding the client through it.

Inquire about specific relevant data. Ask the client questions that are designed to elicit appropriate data, and learn more about the client and his or her concern. The skillful use of questioning can facilitate insight when the probe helps the client recall important data, promotes more comprehensive discussion during the counseling process, and enhances awareness on the part of the client.

Use direct, forceful statements that are designed to encourage the client to think or feel in a new or different manner. This communication process is used when the counselor believes it is necessary to produce some movement on the part of the client. These deliberate statements are designed to upset the client by showing disapproval or rejection of a client's statement, confront the client by sharply pointing out discrepancies, direct the client by strongly advising the client to do something, center the client by not digressing, or self-disclose, which is designed to help the client see things from a different perspective.

Interpret and evaluate various data and information with the client. This process is used when counselors share their knowledge and insight to help clients understand the relationship between various factors. It assists clients in learning more about their strengths and weaknesses, thus providing a deeper understanding of themselves.

SUMMARY

This chapter briefly outlined the person-centered, the Gestalt, and the existential approaches to counseling. The affectively oriented approaches to the counseling process are based on the notions that individuals have the ability to understand themselves, regulate their own behaviors, choose wisely among alternative courses of action, and fulfill their inherent potentials. These strategies stress developing a strong relationship with clients, understanding clients from the clients' own subjective views of life, and dealing with clients in the here-and-now situation. In these strategies the counselor's objective is to have clients improve their ability to deal with their emotions, beliefs, and values; develop more awareness of themselves; and gain more positive feelings of self-worth. Suggestions were made on methods to use to help clients ventilate; obtain catharsis; and gain insight, self-awareness, and self-esteem.

After your comprehensive review of this chapter, you should be able to discuss the major concepts in the three affectively oriented approaches. You should have a clear perception of how and when to employ these strategies. And you should be aware of the processes employed in using these interventions in practice. To develop a high level of competency in applying affectively focused intervention strategies, you will need further study of these approaches and some experience using them under the guidance of an expert.

References

Bohart, A. C. (1980). Towards a cognitive theory of catharsis. *Psychotherapy: Theory, Research and Practice, 17,* 192–201.

Burks, H. M., & Stefflre, B. (1979). *Theories of counseling* (3rd ed.). New York: McGraw-Hill.

Corey, G. (1991). *Theory and practice of counseling and psychotherapy* (4th ed.). Pacific Grove, CA: Brooks/Cole.

Corsini, R. J., & Wedding, D. (Eds.). (1989). *Current psychotherapies* (4th ed.). Itasca, IL: F. E. Peacock.

Frankl, V. E. (1985). *Psychotherapy and existentialism.* New York: Washington Square Press.

Fromm, E. (1955). *The sane society.* New York: Rinehart.

Fromm, E. (1989). *The art of loving.* New York: Harper.

Gilliland, B. E., James, R. K., & Bowman, J. T. (1989). *Theories and strategies in counseling and psychotherapy* (2nd ed.). Englewood Cliffs, NJ: Prentice-Hall.

Gladding, S. T. (1988). *Counseling: A comprehensive profession.* Columbus, OH: Merrill.

Hartmann, G. (1974). *Gestalt psychology: A survey of facts and principles.* Westport, CT: Greenwood.

Ivey, A. E., Ivey, M. B., & Simek-Downing, L. (1987). *Counseling and psychotherapy: Skills, theories and practice* (2nd ed.). Englewood Cliffs, NJ: Prentice-Hall.

Latner, J. (1984). *The Gestalt therapy book.* Highland, NY: Gestalt Journal.

May, R. (1986). *The discovery of being: Writings in existential psychology.* New York: Norton.

May, R. (1989). *The art of counseling.* Nashville: Gardner Press.

May, R., & Yalom, I. (1989). *Existential psychotherapy.* In R. J. Corsini & D. Wedding (Eds.). *Current psychotherapies* (4th ed.) (pp. 363–402). Itasca, IL: F. E. Peacock.

Nichols, M. P., & Efran J. S. (1985). Catharsis in psychotherapy: A new perspective. *Psychotherapy, 22,* 46–58.

Passons, W. R. (1975). *Gestalt approaches in counseling.* New York: Holt, Rinehart & Winston.

Patterson, C. H. (1986). *Theories of counseling and psychotherapy* (4th ed.). New York: Harper & Row.

Perls, F. S. (1988). *Gestalt therapy verbatim.* Highland, NY: Gestalt Journal.

Perls, F. S., Hefferline, R. F., & Goodman, P. (1977). *Gestalt therapy: Excitement and growth in human personality.* New York: Bantam Books.

Polster, E., & Polster, M. (1973). *Gestalt therapy integrated: Contours of theory and practice.* New York: Brunner/Mazel.

Rogers, C. R. (1942). *Counseling and psychotherapy.* Boston: Houghton Mifflin.

Rogers, C. R. (1951). *Client-centered therapy.* Boston: Houghton Mifflin.

Rogers, C. R. (1957). The necessary and sufficient conditions of therapeutic personality change. *Journal of Consulting Psychology, 21,* 95–103.

Rogers, C. R. (1972). *On becoming a person.* Boston: Houghton Mifflin.

Scheff, T. J. (1979). *Catharsis in healing, ritual, and drama.* Berkeley: University of California Press.

Shilling, L. E. (1984). *Perspectives on counseling theories.* Englewood Cliffs, NJ: Prentice-Hall.

Van Kaam, A. (1966). *The art of existential counseling.* Wilkes Barre, PA: Dimension Books.

Van de Reit, V., Korb, M. P., & Gorrell, J. J. (1980). *Gestalt therapy: An introduction.* Elmsford, NY: Pergamon Press.

Yalom, I. D. (1981). *Existential psychotherapy.* New York: Basic Books.

6 $\parallel$ *Performance Focused Counseling Strategies*

INTRODUCTION

This chapter presents the information necessary to understand the behavioral or performance approach to helping clients. The techniques outlined in this chapter are seen as ways of helping clients learn to modify their activities and thus are considered educational in nature. You should be able to respond to the following discussion questions after studying this chapter:

- Indicate the differences among the three major behavioral approaches to counseling.
- Identify the important steps in the behavioral counseling process.
- Specify the elements of a contingency contract.
- Name the major techniques used in the social learning approach and indicate when they might be useful in a counseling situation.
- Describe how positive and negative reinforcements and punishments can be used to change behavior.
- Discuss some key behavioral techniques such as assertiveness training, aversion techniques, flooding, paradoxical intention, and systematic desensitization.

This chapter discusses ways to work with clients from a performance or behavioral point of view. The basic premise in this approach is that individuals develop their personalities or characteristic ways of behaving as a result of their life experiences, and they have *learned* to behave or act the way they do. Some of this behavior is appropriate and desirable, whereas other behavior can be inappropriate and undesirable. Individuals who manifest maladaptive behavior have failed to learn correctly, and the way to help these people is to teach them appropriate ways to behave. Furthermore, because all aspects of a person's life are influenced by how that person acts, helping an individual change a behavior often helps that person modify other dimensions of his or her life in significant ways. Performance focused interventions are those that help clients increase the likelihood of performance of an existing activity, initiate a new activity, or eliminate or decrease the likelihood of performance of an activity.

This chapter contains five sections. First, an overview of the essential steps used in the behavioral counseling approach are outlined. Second, the meaning and use of contingency contracts are explained. Third, the basic principles of observational and simulated learning are delineated. Fourth, the major concepts of contingency management and operant conditioning are presented. And, finally, the essential ideas of stimulus response and classical conditioning are specified. In each section, suggestions are offered for using these intervention strategies in the counseling process.

BEHAVIORAL COUNSELING

The term *behavioral counseling* can be defined as a learning process by which the counselor helps the client learn something new or learn how to change a specific behavior in a particular situation. Behavioral counseling emphasizes identification of the client's problem, clearly labeling the problem in behavioral terms, defining the desired behavioral goals or changes, specifying a plan to bring about the desired changes, and observing and recording the client's behaviors. The counselor takes an active role in teaching the client new ways of behaving and involves the client as much as possible. Typically, after a counseling relationship has been established and the problem has been identified, the behavioral counselor follows these five sequential steps.

1. Describe the problem behavior and the context in which it occurs. Identify the items in the behavioral formula ($S^D \rightarrow R \rightarrow S^R$) where S^D is the discriminative stimulus or antecedent or situation that occurs prior to the behavior; R is the problem behavior or response to the discriminative stimulus; and S^R is the consequence or behavior that occurs after the problem behavior and thus reinforces the problem behavior. It is helpful to find answers to the following questions:
   a. What stimuli or events occurred before the target behavior?
   b. What was the biological and psychological state of the individual when the behavior occurred?
   c. What is the problem behavior or the response to the prior events?
   d. What is the nature of the relationship between the problem and the reinforcers (S^R)?
   e. What types of reinforcers are present?
2. Identify the observable and measurable components of the problem behavior.
   a. Note the *frequency* of occurrence or number of times the response occurred in a period of time; the *intensity* or forcefulness of the response; the *duration*; and the *topology* or form or shape of the behavior.
   b. Collect these data by means of the continuous, time-interval or situation sampling method and record it in a way that is relatively simple,

portable, and economical. Suggested methods include using note-books, logs, diaries, mechanical devices, tape recorders, charts, and graphs. Martin and Pear (1983) provide more details on appropriate recording procedures.

3. Specify the desired goal. Indicate whether the behavioral goal is:
   a. *response acquisition,* or learning new skills or behaviors;
   b. *response facilitation,* or learning to use presently existing skills more frequently or in a variety of situations; or
   c. *response inhibition,* or learning to decrease the likelihood of un-desirable responses.

4. Agree on the set of standards or performance criteria that will be achieved.
   a. Identify what needs to be modified in the behavioral formula $(S^D \rightarrow R \rightarrow S^R)$.
   b. State the environment where the behavior should be performed or the most favorable discriminative stimuli (S^D).
   c. Indicate the level of performance or criterion level in terms of fre-quency, intensity, or duration.
   d. State what reinforcements will be used to bring about the new behavior.

5. Develop a plan for the behavior changes.
   a. State the stimulus-response conditioning, simulation, or operant con-ditioning techniques that will be employed.
   b. Indicate who will be responsible for what activities.
   c. Identify the consequences for meeting or not meeting targeted goals and how they will be provided.

6. Describe to the client the methodology planned, the reason for each activity, and the responsibilities of the client and the counselor.

7. Implement the plan. During this treatment phase, continued observa-tions should be made on the frequency, duration, and intensity of the target behavior in order to evaluate the client's progress and the factors that facilitate or impede the client's development.

8. As the client manifests the appropriate behavioral changes, every ef-fort must be made to develop variable schedules of reinforcement and to practice the new behaviors in realistic settings.

9. When the new behaviors become established, plans should be made to terminate counseling and methods should be established for client follow up.

CONTINGENCY CONTRACTS

Counselors using a behavioral approach to counseling often employ a technique known as the *contingency contract.* This is also referred to as a behavioral or psychological contract. Contracts indicate a mutual agreement to work toward a specific goal or to exchange services or goods. The contract normally specifies who is responsible for what and under what conditions, and what contingencies apply if the contract is broken. Contracts form the basis of much of our society. We often have explicit written contracts for legal activities such as buying property, opening bank and brokerage accounts, and marketing our personal services. On the other hand, much of our social interaction is made by implicit verbal agreements, in which the contingencies of breaking the contract are not determined; for example, when someone is late for a dinner engagement, the consequences of that behavior are often not spelled out in advance. Behavioral or contingency contracts are similar to our other social contracts. However, for the sake of clarity and complete understanding, they should be explicit and should specify the following points:

1. Outline a realistic, clear, and detailed description of the client's goals. These goals can be stated in terms of a range of behaviors, such as "I will compliment my neighbor at least once a day," and "I will try to compliment him three times a day."

2. State the time and frequency by which the goal or behavior is to be measured—for example, "Within three weeks I will visit my mother-in-law three times."

3. Indicate the contingencies or the consequences of meeting or not meeting the goal. Specify the positive or negative consequences that will be given, when and where they will be given, and who will give them. Articulate how these are contingent upon fulfilling or not fulfilling the time and frequency goals. These rewards and punishments should be minimal, immediate, and frequent—for example, "You will earn one dollar each time you do this," and "You will forfeit two dollars when you do that more than twice a day." Be careful not to agree to consequences that are too intense and thus unrealistic; for example, a consequence of $100 is too large for most clients.

4. Specify the means by which contract terms are recorded, measured, or observed and how these terms and observations will be reviewed; for example, the person who is trying to stop smoking may be instructed to keep a daily log of the time, place, and events that preceded his or her smoking a cigarette or having a desire for one.

5. Attach a bonus clause to the contingency contract to indicate additional rewards that can be earned if and when the minimal goals are exceeded; for example, the client who is trying to learn to get his or her homework done on time may be given the target goal of finishing the work by 9 p.m. every day and given the reward of watching television from 9 p.m. to 10 p.m. If the client finishes by 8:30 p.m., he or she may be awarded the pleasure of watching an extra half hour of television.

OBSERVATIONAL AND SIMULATED LEARNING

Basic Concept

The first important behavioral counseling approach discussed is based on the concepts of observational and simulated learning. Observational and simulated learning stresses that individuals learn new ways of behaving by observing others behave in certain ways and by practicing behaviors in simulated ways. The principles underlying this approach to learning have been articulated by Bandura and Walters (1963). Individuals can learn through this social learning process in two distinctive ways: by observation and by simulated experiences. In the observational learning approach, some individual or group acts as a model or a stimulus for changing the thoughts, feelings, or behaviors of the client. The client observes the model's performance and, without actually engaging in the behavior, may obtain new knowledge, modify attitudes, and acquire new behavior. In the simulation learning approach, the client plays a role, or simulates himself or herself in a situation or relationship. This behavioral practice enables clients to overcome inhibitions, deal with their anxieties, and learn appropriate skills. It is common to use both observational and simulated approaches in a sequential process.

Observational Learning

Observational learning is a common occurrence in life—people learn by observing the behavior of others. Informally, this learning process is used throughout our daily lives, as evidenced by the influence others have on our dress codes and our moral and ethical behavior. In a more formal setting, this learning process is used by mentors, teachers, and instructors who model certain skills and expect their students to spend hours learning to copy their actions. The observational learning process is also referred to as modeling, imitation, copying, and mimicry. This learning process has two phases: the *acquisition* phase, which occurs when the observer attends, watches, and listens to the model; and the *performance* phase, which occurs when the observer performs the desired activity.

Modeling procedures can be overt, symbolic, or covert. *Overt* behaviors are activities that are performed in the here-and-now situation. Overt or participant modeling occurs when a live model performs the behavior in the presence of the observer—for example, when the counseling instructor demonstrates how to open a counseling session. This method has the advantage of allowing the observer to identify with the model and see actions that the observer wants to emulate. The disadvantage is that the model's actions or behaviors cannot be controlled; thus, clients may see undesirable behaviors as well as desirable ones.

Symbolic behaviors take place in an abstract or symbolic way. The symbolic modeling procedure is being used when the target behavior is presented through the media—for example, when students who are studying Rogerian counseling view a film of Rogers counseling a client. This technique has the advantage of allowing the counselor to have control over the modeled behaviors; the counselor can select relevant portions of the target behavior, and replications are easy to do. But the observer cannot identify personally with the model, and relevant symbolic models who demonstrate the desired performance may not be available or may be difficult to find.

Covert behaviors take place in our thoughts. Covert modeling procedures are employed when the model is presented through the imagination—for example, when the instructor asks you to imagine yourself in a particular situation or in a role-playing exercise. This approach has the advantage of having complete control of the target behavior because one's imagination can supply relevant materials. It has the disadvantage of the possible loss of reality.

The modeling process can have one of three major effects on the learner: the observer learning effect, the response facilitative effect, or the disinhibiting effect. The *observer learning effect* occurs when a person learns something by observation alone. Because human beings learn simply by seeing things, clients can learn new behaviors solely by observing a model's behavior; for example, by seeing someone bake a cake or solve a problem, an individual can learn how to bake a cake or solve problems. The *response facilitative effect* occurs when an individual sees a model do something that he or she knows how to do and the observer then does it, providing there are no constraints. In other words, the model prompts the behavior; for example, one person may light a cigarette after another person has lit one. In the *disinhibiting* or *inhibiting effect,* an individual sees something done that he or she knows how to do but that person is either disinhibited from repeating the behavior (for example, the player who carries out a particular play after seeing the coach demonstrate the play) or inhibited from repeating the behavior (for example, the person who does not buy an airplane ticket after seeing or reading about a serious airplane accident).

Observational or imitative learning is fostered by using the following eight-step process:

1. Assist the client in identifying the behavior or skill that he or she wants to learn.
2. Eliminate or minimize distracting stimuli and intensify relevant stimuli. This may be done by helping the observer relax.
3. Help the observer focus on what to look for in the model by giving a set of instructions or relevant questions.
4. Select a warm, nurturing model whom the client can respect and identify with. Identification can be enhanced by matching the observer with the model in terms of competence, age, gender, and cultural characteristics. However, the model should be sufficiently distant from the observer in age or competency so that the observer will respect the model and be motivated to learn.
5. Break down any complicated behavior into small incremental steps that can be modeled and observed.
6. Observe the model and discuss the observed behavior with the client. Observe the behavior in order from the least difficult responses to the most difficult. The client should carefully observe the antecedents and the consequences of the behavior.
7. Have the client practice the new behavior or skill as soon after the observation stage as is practical. Provide support and verbal encouragement. This practice should occur both under the direction of the counselor, who can guide the practice, and as homework, so the practice is rehearsed.

8. Evaluate the procedure with the client. If the desired skill was obtained, the intervention was successful. If the desired skill was not obtained, further modeling or assistance with other interventions may be called for.

Simulated Learning

Simulated learning has proved to be an effective learning process. It is a common occurrence in our everyday experiences. Skiers who practice in their basements and job applicants who mentally prepare for an interview are using this learning process. Simulations can deal with realistic or exaggerated situations; overt events or covert thoughts; or past, present, or future occurrences. Simulations are sometimes referred to as replications, practice, or drill.

Three major forms of simulations exist: role-playing, role reversal, and dialoguing. *Role-playing* is used when the client plays himself or herself in a particular situation or relationship. It has the advantage of allowing the client to practice being in a certain situation. For example, a placement counselor may have a client play himself in a mock job interview so the client can experience what a job interview is like. *Role reversal* is employed when the client plays someone else in the situation. This has the advantage of having the client experience how another person may react in a given scenario. For example, the placement counselor may have the client play the personnel manager in a mock job interview so the client can gain an understanding of what the employer may be looking for in a job applicant. *Dialoguing* is the process used when the client plays himself or herself and someone else at the same time and conducts a dialogue between the two. For example, the placement counselor can have the client play himself or herself and the personnel manager during the mock job interview. This has the advantage of having the client experience both sides of an issue.

Learning through the simulation process is facilitated by the following five-step procedure:

1. Help the client specify the target behavior, attitude, or performance to be learned. Be as specific and concrete as you can.
2. Encourage the client to determine the environment or situation where the skill needs to be used, the fear reduced, or the attitude changed.
3. Arrange the simulations in a hierarchical order; start with small scenes, minimal interchanges, least difficult or least anxiety-provoking situations, and plan a gradual expansion to more difficult, more complex scenes. Involve the client in planning the hierarchy.
4. Apply the hierarchy. Begin with the first simulation using covert methods and practice. Have the client discuss the experience, and provide feedback and verbal reinforcement. Move on to more complex situations and progress from minimal-risk to minor-risk situations. Provide ample practice, vary simulations, and give the client plenty of feedback and reinforcement.

5. Have the client apply the new behavior or skill in a real situation. It is important to review the experience with the client to give the client a chance to reflect on this experience and gain some insight into his or her behaviors. If the trial was successful, encourage the client to repeat the experience several times while providing support and feedback. If the experience was of limited success, repeat the simulated experience until more mastery and comfort are assured. If the experience was not successful, repeat the simulation and investigate the possibility of incorporating other techniques into the intervention program.

Applications and Advantages of Observational and Simulated Approaches

Modeling and simulated learning approaches are extremely useful in a variety of counseling situations including educational and vocational exploration; decision-making interventions; catharsis, insight, and self-awareness concerns; and modification of social behaviors. The counseling process, which relies on verbal interchanges, lends itself to observational learning and simulations. The counselor has considerable control in this approach through the selection of appropriate situations, the amount of repetition, and the intensity of the experience. This technique can be used in a systematic manner. The process is fun and engaging for clients.

CONTINGENCY MANAGEMENT

Basic Concept

The second major behavioral counseling approach is contingency management. *Contingency management* is the name given to those behavioral counseling strategies based on the concept that the responses that follow the behavior are the behavioral modifiers. This can be represented by the second part of the behavioral formula ($R \rightarrow S^R$) and is often referred to as instrumental, operant, or type-R conditioning. This learning method can be stated as the *principle of reinforcement.* This principle, sometimes called the *law of effect,* can be stated in three ways:

- behavior is controlled by its consequences;
- behavior that is followed by a satisfying state of affairs is strengthened or stamped in, whereas behavior that is followed by an annoying state of affairs is weakened or eliminated; or
- acts that are rewarded are enhanced, and acts that are punished are curtailed.

This conditioning can be exemplified by the student who studies for an exam. If he gets a good grade, his study behavior is reinforced and likely to be repeated. If he fails his exam, he is apt to feel his studying did not pay off and hence may be less likely to study for an exam in the future. Another example is the child who puts her hand in the cookie jar. If she gets a cookie, her behavior is reinforced. If she doesn't find a cookie, her behavior is not reinforced and is not likely to be repeated in the near future. Skinner (1938, 1953, 1989) made significant contributions to our understanding of these behavioral principles; and Kanfer and Goldstein (1986), Kazdin (1989), and Krumboltz and Thoresen (1976) provide excellent examples of how these principles can be used in a variety of settings.

Before discussing how contingency management techniques can be used to increase or decrease the likelihood of a behavior, it is important to understand the following terms: operants, behavioral consequences, primary and secondary consequences, reinforcement schedules, aversive stimuli, the effectiveness of operant contingencies, and the meaning of other operant terms.

OPERANTS.   Operants are the behaviors or the responses (R) that precede and are related to the reinforcing stimulus (S^R). In the previous examples the behavior (R) "studying" was reinforced by the stimulus (S^R) "the good mark on the test"; and the behavior "putting the hand in the cookie jar" was reinforced by the cookie. Activities that operate on or have an effect on our environment are considered to be operant behaviors. Thus, most human behaviors, such as walking and speaking, are considered to be operants.

Operant behaviors have three major characteristics. First, operants are voluntary or freely emitted behaviors that operate in and on the environment; thus, they are controlled only by their environmental consequences. Theoretically, there is no pairing of the prior stimulus and the response. The contingent reinforcer is in control, and the instrumental response is influenced by its outcome. For example, the behavior of studying is controlled by the result of studying, or the mark on the exam, and putting the hand in the cookie jar is controlled by the reward of getting or not getting a cookie. Second, operants are learned behaviors. These behaviors are distinct from respondent behaviors, which are usually innate or maturational. For example, a knee jerk is an innate, respondent behavior; studying and putting one's hand in a cookie jar are learned behaviors. Third, operants have a contingent relationship to the environmental events that follow. To be in a *contingent relationship,* event B may follow event A, but need not do so. For example, if you go to the beach, you may get wet. In contrast, to be in a *dependent relationship,* event B must necessarily follow event A. For example, if you swim, you will get wet. Both "studying" and "putting one's hand in the cookie jar" are operant behaviors that may be followed by different consequences. Although the term *free operant* is often used to describe the absence of constraints, human behavior almost always occurs in a context that sets some boundaries; hence, there is usually an interdependence between the behavior (R) and the prior events (S^D) or the environment.

BEHAVIORAL CONSEQUENCES.   Operant behavior can result in behavioral consequences or operant contingencies that may be pleasant or unpleasant; they may occur or be given, or they may not occur or be removed. It is important to understand how these four contingencies can be employed to help increase or decrease the likelihood of a behavior.

When the consequence or contingent response tends to *maintain* or *increase* the probability of that operant behavior, then *reinforcement has occurred.* Reinforcement can be positive or negative. Reinforcement is positive when a pleasant consequence follows or is added or given—for example, when a pleasant feeling follows doing a favor for a neighbor or when a gold star is given. Reinforcement is negative when an unpleasant consequence disappears or is removed or taken away—for example, when a pain goes away after medication or when a detention is removed. In all these cases, reinforcement of the desired behavior has occurred.

When the consequence or contingent response tends to *extinguish* or *decrease* the probability of that operant behavior, *punishment has occurred.* As with reinforcement, punishment can be positive or negative. Punishment is positive when an unpleasant consequence follows or is added—for example, when a student fails a test for not studying or when detention is assigned. The punishment is negative when a pleasant consequence goes away or is removed—for example, when the gold star is taken away or when the cookie jar is deliberately left empty. In either case the target behavior is less likely to occur in the future. Four possible consequences or operant contingencies are illustrated in Table 6-1.

Another way to portray these four contingencies is by the following outline:

1. Reinforcement is used to start or increase a behavior.
   a. It is positive when something pleasant occurs or is given.
   b. It is negative when something aversive is removed or disappears.
2. Punishment is used to decrease or eliminate a behavior.
   a. It is positive when an aversive result occurs or is given.
   b. It is negative when something pleasant is taken away or lost.

REWARDS. Rewards are not always reinforcers. A reward is considered to be something given to a person for completing a particular activity—for example, giving a student a candy bar for getting 100 percent on a test. However, a reward may not be valued by the person and hence not really serve as a consequence for increasing the target behavior. The student may not like candy, or he or she may feel it is a bribe and not at all related to the activity of studying. Rather than using a reward, it is more desirable to have the person see the consequence as a logical and natural one. For example, instead of using a candy bar as a reward

T A B L E 6-1
The Four Possible Operant Contingencies

	Added = Positive	*Removed = Negative*
Pleasant consequence	Positive reinforcement	Negative punishment
Unpleasant consequence or aversive stimulus	Positive punishment	Negative reinforcement

for going into a pool on a hot day, the natural consequence of the cooling of the body should be allowed to serve as the reinforcer. Rewards, however, can and are used as reinforcers. To be useful, they must be valued by the persons concerned and applied in a consistent fashion.

PRIMARY AND CONDITIONED CONSEQUENCES. Consequences can be primary or conditioned. A *primary* or *universal reinforcer* is a consequence that is related to maintaining and perpetualizing life or the satisfaction of physiological needs. Examples of primary reinforcers include food, water, a balanced body temperature, rest and sleep, procreation, and breathing. A punishment is considered primary and universal when the consequences have a universally aversive reaction. Examples of aversive stimuli include loud noise, bright lights, temperature extremes, loss of sleep, and other painful situations.

A *conditioned consequence* is a stimulus or an object that initially had no value but that by continued pairing with a primary consequence has assumed reinforcing or punishing qualities. Conditioned positive reinforcers include verbal praise such as "That was a nice thing to do," nonverbal gestures like smiles and winks, and symbolic or token objects such as money or gold stars. Conditioned positive punishments include verbal rebukes such as "Don't do that!" or nonverbal gestures like frowns and stares or symbolic objects such as demerits. Conditioned consequences are used when we want to reinforce desired behaviors and punish undesired ones.

Attention is a powerful reinforcer because it meets a basic human need. It can serve to reinforce both appropriate and inappropriate behaviors. Thus, a child whose ill-advised behavior gains attention is likely to repeat that behavior because it was reinforced. Similarly, withdrawing attention is an excellent form of negative punishment because it removes something desirable. Thus, the child who engages in an ill-advised behavior such as a temper tantrum should not be given the attention he or she wants.

REINFORCEMENT SCHEDULES. For new behaviors to become established, they do not have to be reinforced every time they occur. Reinforcement can be scheduled at different times and rates. The major kinds of reinforcement schedules are continuous, or uninterrupted, and intermittent, or periodic. When trying to establish a new behavior, it may appear that constant reinforcement is desirable to encourage the behavior to become habitual; however, the fact is that the exact opposite is true. This principle is known as *Humphrey's paradox.* Thus, in order for a behavior to become "stamped in," it is desirable to start with a continuous reinforcement schedule and move to an intermittent schedule with longer gaps between reinforcers.

Continuous reinforcement (CR) occurs when a behavior is reinforced every time it occurs. When one wants to develop or establish a new behavior, it is a very effective schedule. However, when the behavior reaches a plateau (called the *asymptotic level*), continuous reinforcement is no longer effective. Thus, one must learn to reinforce on an intermittent basis. For example, when one wants to establish a behavior such as doing homework, cleaning a room, or practicing the piano, the behavior initially should be reinforced each time it is done by praise or some token reinforcer like a gold star. However, as soon as feasible, the reinforcement schedule should be switched to an intermittent one, which lessens the density or frequency of the reinforcement gradually and progressively.

Intermittent reinforcement (IR) occurs when the behavior is reinforced on a periodic basis. Reinforcing a behavior periodically increases the likelihood that the behavior will be repeated. The reinforcement can be scheduled to occur on a time (interval) basis or an event (ratio) basis and can occur at fixed or variable points or places. There are four possible ways to reinforce on an intermittent basis: the fixed-ratio schedule, the variable-ratio schedule, the fixed-interval schedule, and the variable-interval schedule.

The two *ratio schedules* require that a certain number of responses be given before the reinforcement can be administered. Ratio schedules are similar to piecework, in which one is paid for accomplishing a certain task. This form of reinforcement tends to increase the pace of the target behaviors, because the more rapidly one accomplishes a task, the sooner a reinforcement occurs. The *fixed-ratio* schedule (FR) is a pattern of reinforcement that is presented on a regular-event basis; for example, one might give written compliments to a student for every fifth problem or every tenth page of work. This schedule is associated with a tendency for the person to slow down after the reinforcement is given. The *variable-ratio* schedule (VR) is a reinforcement pattern that is constructed to present the reinforcement on a variable-event basis; for example, one may develop a reinforcement system to give five gold stars for a term paper but vary the events that will be reinforced. The schedule may grant the star after the 1st, 8th, 12th, 20th, and 22nd pages. Playing slot machines follows a variable-ratio reinforcement schedule. Variable-ratio schedules are not effective when reinforcement is too spread out.

The two *interval schedules* require that a certain amount of time lapse or pass before reinforcement is given. Interval schedules are preferred by many behaviorists because this type of schedule avoids having people rush to accomplish a task. The *fixed-interval* schedule (FI) is a pattern of reinforcement that occurs at specific, or fixed, time intervals; for example, one might use verbal praise as a reinforcer and compliment the individual every 10 minutes. The *variable-interval* schedule (VI) is a reinforcement pattern that is based on a variable time interval. For example, one might develop a token reinforcement system that gives a gold star five times during the day and decide to vary the time that the reinforcer is given, such as after 1 hour, 3 hours, 4 hours, 5 hours, and 8 hours. The variable-interval schedule explains why it is sometimes hard to extinguish an undesired behavior. If an "acting-out" behavior occasionally gains a child attention, that behavior is being reinforced, and thus it is likely to be maintained.

A VERSIVE STIMULI OR UNPLEASANT CONSEQUENCES. In helping clients modify their behaviors, it is often desirable to assign an unpleasant consequence (positive punishment) or to remove an aversive stimulus (negative reinforcement). The use of severe primary aversive stimuli such as loud noises, extreme temperatures, and intense pain, raises serious ethical issues and is justified only when less aversive means have not or cannot work and when the maladaptive behavior is serious enough to warrant the use of these extreme means. Conditioned aversive stimuli such as verbal rebukes and minor primary aversive stimuli such as slight shocks are employed when aversive methods are appropriate.

It should be noted that a contingent stimulus can be pleasant or aversive, depending on its effect on a particular behavior, and a particular behavior may have aversive and pleasant consequences simultaneously. For example, when a student misbehaves in class, other students may cheer (the pleasant consequence), while the teacher scolds (the aversive consequence); and when the team commits an error, the home fans boo (an aversive consequence) while visiting fans cheer (a pleasant consequence). Careful thought must be given to select consequences that are truly reinforcing when one wants to increase the likelihood of a behavior and truly punishing when one wants to decrease or eliminate a behavior.

Aversive consequences can also have undesirable side effects. These side effects occur when an individual learns to deal with the unpleasant consequences in one or more of the following three ways: operant escape, operant avoidance, and aggression. *Operant escape* occurs when the individual removes himself or herself from the aversive or painful experience; for example, when a father is yelling at his son, the son can try to escape the unpleasant event by leaving the room (physical escape) or by daydreaming (mental escape). *Operant avoidance* occurs when the individual acts to avoid a painful result or experience; for example, when a student believes that taking a particular class will result in failure or boredom, the student will try to avoid the class. *Aggression* occurs when the individual reacts to an aversive stimulus with aggression; for example, when a teacher scolds a student, the student sasses or physically rebuffs the teacher.

EFFECTIVENESS OF REINFORCEMENT AND PUNISHMENT. Reinforcement and punishment are most effective when the consequences are (1) contingent upon the behavior, (2) consistently used for the same behavior, (3) immediately used with minimal time delays, (4) intense enough to be seen as a consequence of the behavior and yet weak enough to allow for an increase in the intensity of the consequences in the future, and (5) appropriate to the behavior. Furthermore, punishment is most effective when combined with positive reinforcement of alternating behaviors and when chances for escape, avoidance, and aggression are minimized.

OTHER OPERANT TERMS. Several other terms are used when behavioral and contingency management methods are employed. These terms are response differentiation, shaping, chaining, prompting and fading, satiation, deprivation, restraint, and the Premack principle.

Response differentiation is the process of reinforcing one level or variant of an existing behavior while punishing or withdrawing reinforcement from other levels or variants of the behavior; for example, a parent who is trying to help a child lose weight may compliment the child every time he or she eats a desirable food and frown or show no reaction to the child when he or she eats an inappropriate food.

Shaping is the process of reinforcing successive approximations to a target behavior until the new behavior has been learned. The terminal goal is broken down into a series of steps, and each slight change is reinforced. The size of each step and its intensity must be determined on an individual basis. For example, in learning how to reflect the content and feelings of a client statement, the behavior might be broken down into the following sequences: a response that mirrors the content of a client's statement, a response that mirrors the feeling of a client's statement, and a response that goes beyond the surface of the client's statement and reflects the underlying content and feeling. In training someone to learn this behavior, it is possible to shape the responses by systematically teaching the sequence and reinforcing appropriate student responses at each level.

Chaining is the process of combining previously learned behaviors into a more complex behavior. Each event or activity has a triple function: it is a behavior, it is the discriminative stimulus for the behavior that follows, and it is the reinforcement stimulus for the prior behavior. The behaviors "walk to the office," "sit down," "talk to the counselor," "relax," and "leave the office" can be labeled as the discriminative stimulus S^D, the behavior R, or the reinforcing stimulus S^R, depending on which link of the chain of human behavior is being analyzed. Three links of a chain of human behavior are represented in Table 6-2.

T A B L E 6-2
Links in a Chain of Human Behavior

Discrimination Stimuli	Behaviors	Contingent Stimuli
1. walk to office (S_1^D) →	sit down (R_1) →	talk to counselor (S_1)
2. sit down (S_2^D) →	talk to counselor (R_2) →	relax (S_2)
3. talk to counselor (S_3^D) →	relax (R_3) →	leave office (S_3^R)

In learning a new complex behavior, a chain of simple behaviors is reinforced until the complex behavior is learned.

Prompting is the act of providing a stimulus or a cue for a person to act in a certain way (for example, when mother says, "Set the table"). In starting a new behavior, a conspicuous prompt is often necessary. The prompt should be gradually removed so that the new behavior is eventually done without a hint or a cue. The process of gradually removing the prompts is called *fading*—for example, to teach a child to set the table, a prompt such as "John, set the table" can be used. Because it is most desirable for the child to set the table without the prompt, the parent would gradually remove and not offer the cue.

Satiation has occurred when too much reinforcement of one type has been given. If an individual has too much of a good thing, the reinforcer loses its power to reinforce the behavior; for example, a student who has recently had a lot of candy at a birthday party is probably satiated and will not see a piece of candy as an incentive to act in a particular way. Sometimes satiation causes other actions; for example, when a person praises others continuously, his or her sincerity may be questioned.

Deprivation is the term used when no reinforcement has been provided. This deprivation of reinforcement normally increases the reinforcing power of a stimulus. Thus, if a child likes candy but has not had any in a long time, the child has been deprived of candy as a reinforcer and the candy can probably serve as a strong incentive to act in a particular way. Deprivation of important reinforcers can cause severe problems; for example, students who receive no praise for their work can be damaged or turned off from a subject.

Restraint is the word employed when physical force is used to prevent a behavior from occurring. This method of behavioral change implies an authoritarian approach and is best used when the client's well-being or some other person's well-being suffers as a result of the behavior; for example, a mentally retarded child may be prevented from hitting his or her own head by restraining the movement of the child's arms.

The Premack principle is the premise used when we attempt to increase low-probability behaviors by making them contingent upon high-probability behaviors; for example, "You can watch TV (the high-probability behavior) after you do your homework" (the low-probability behavior). The term was named after the psychologist Dave Premack (1965), who wrote about its applications in behavioral techniques. This principle is also referred to as grandma's rule.

Increasing the Likelihood of a Behavior

Contingency management methods can be used to increase the likelihood of a behavior. Because the operant definition of reinforcement is the procedure by which the operant behaviors are increased or maintained, the way to help clients increase the probability of a particular behavior is by positive and negative reinforcement methods. Positive reinforcement methods provide a pleasant consequence as the result of the behavior. Positive reinforcement is a normal and consistent process throughout our entire lives. Some typical examples are good grades and a sense of accomplishment from studying, verbal praise for doing something helpful for others, and the salary check or the token reinforcer for performing a work-related task.

Negative reinforcement methods remove aversive consequences as the result of the behavior. Removing aversive stimuli is usually less desirable than positive reinforcement, not because it does not work but because the aversive stimuli may not be present (student is not in detention, pain is not present) or because the use of aversive stimuli has some undesirable side effects. Some common examples of how negative reinforcement is used in our lives include removing anxiety by studying, removing pain by going to a physician, and removing teasing when we do what others want us to do.

The following steps outline a systematic way to help clients increase and maintain the likelihood of a particular behavior using operant conditioning methods:

1. Help the client identify the target behavior or performance to be changed. Be as specific and concrete as possible. If necessary, break down complicated behaviors into the links of a chain.

2. Analyze the behavioral chain and determine what current reinforcers, if any, exist for the target behavior and notice what reinforcement schedule is in effect.

3. Identify reinforcers for the target behavior. Determine what pleasant consequences can be added and what unpleasant ones can be removed. Many behavioral patterns of individuals are the result of poorly arranged or poorly thought-out reinforcers. It is therefore important to analyze the consequences of the client's present behaviors in terms of the type of reinforcer as well as the frequency, intensity, and duration of these reinforcers. What changes might be made to this pattern?

4. Look into the possibility that the client has experienced saturation or deprivation of some reinforcers. If either situation exists, can these reinforcers be avoided or used to increase behavior?

5. Is a high level of behavior present? If so, can the Premack principle be used to increase the likelihood of the lower level behavior by making the high-level behavior contingent on the lower level behavior?

6. Identify a schedule of reinforcement. Note the frequency and the duration of the reinforcement pattern and specify by whom, how, and when the reinforcement will be administered.

7. If the targeted behavior is complex, break it down into incremental sequences and plan to shape the desired behavior.

8. Encourage the client to implement the program. If necessary, combine with observational and simulated learning experiences to help the client.

9. Provide positive feedback and evaluate the client's progress step-by-step through the behavioral change program.

10. Continue the conditioning until the behavior is increased and maintained for a reasonable period.

11. Plan client follow-up to see if the targeted behavior has been maintained.

Decreasing the Likelihood of a Behavior

Operant conditioning methods may be used to decrease or eliminate the likelihood of a behavior. Because punishment is defined as a contingent stimulus that decreases a behavior, the contingent methods used to decrease or eliminate a behavior are positive and negative punishment.

Positive punishment methods present an unpleasant consequence as a result of engaging in a particular behavior. The use of unpleasant consequences or aversive stimuli is a common phenomenon in our society. When an individual engages in an undesirable behavior, that individual is subject to some type of undesirable consequence. Children who misbehave get scolded, and the team member who plays poorly gets booed. Punishment used effectively can stop an unwanted behavior quickly, facilitate discriminatory learning, and serve as an instructional example to observers.

Punishment can have three bad side effects. First, it can lead to the use of escape, avoidance, and aggression. Second, it can provide a poor social learning model; for example, when the teacher scolded a student, Jim, he and his peers learned to scold and they in turn scolded Jim's younger sister. And third, the use of primary aversive stimuli can have severe physiological side effects; thus, it can do more harm than good.

Negative punishment methods remove pleasant consequences as a result of engaging in a particular behavior. Removing pleasant consequences is an effective tool in the learning process. In order to use it, of course, the pleasant consequences must be present. Negative punishment can be given in one of three ways: time-out, response cost, and extinction. *Time-out* occurs when the possibility for all reinforcement is removed or all pleasant consequences are cut off for a time—for example, when a player is removed from a game. *Response cost* occurs when part or all of a pleasant consequence previously earned is removed; for example, when one receives a fine for passing a red light or for some other infraction, the previously earned money is the pleasant consequence that is removed. *Extinction* occurs when consequences are discontinued or withdrawn completely, and neutral consequences or events are presented where positive outcomes used to exist—for example, deliberately keeping the cookie jar empty or ignoring a loud child when loudness had previously obtained attention. The use of this form of negative punishment often increases the behavior rather dramatically before the new consequences become evident.

NATURAL AND LOGICAL CONSEQUENCES. Whenever possible, it is desirable to allow the natural or logical consequences of an ill-advised behavior to serve as the punishment. *Natural consequences* are those that result from a behavior without the intervention of another person. These are the universal, inevitable consequences that are true for all human beings regardless of ethnic background or social class. Examples include "If I forget my lunch money, I can't buy lunch"; "If I oversleep, I will be late"; and "If I don't take care of my laundry, I will not have clean clothes to wear."

Logical consequences are the results that follow logically from an ill-advised behavior. These consequences require the intervention of another person or the application of a rule or law. Social conventions, mores, and group expectations usually require certain behaviors and built-in reinforcement procedures. These consequences might be avoided, but another person or group is normally present to impose the sanctions. Examples include "If I go to school late, I will get detention"; "He who breaks the window, fixes it"; and "If I don't show up Thursday, the gang will ignore me."

The following steps indicate a procedure that can be used to help clients decrease or eliminate the likelihood of a particular behavior using type-R conditioning:

1. Help the client identify the behavior to be decreased. If the behavior is complicated, break it down into links of a chain. Help the client be as specific as possible.
2. Analyze the behavioral chain to determine what contingencies presently exist for the target behavior. Note all the positive and negative reinforcers for the current behavior. How strong are these reinforcers?
3. Identify ways to punish the target behavior. What natural and logical consequences are available to help decrease the ill-advised behavior? Determine what aversive stimuli can be given or positive consequences removed. Specify who will administer the punishment and how and where the punishment will be administered.
4. Look into the possibility of initiating or strengthening a desired behavior. It is frequently wise to increase one behavior while decreasing another one.
5. Is it possible to eliminate the current reinforcers by satiation? If so, what are the consequences?
6. Encourage the client to implement the program. If necessary, combine with other techniques, such as observational and simulated experiences, to lessen the likelihood of the behavior.
7. Provide positive feedback and evaluate the client's progress step-by-step through the program.
8. Continue the conditioning until the targeted behavior is decreased or eliminated for a reasonable period.
9. Plan to follow up with the client periodically to see if the desired behavioral change has occurred.

CLASSICAL CONDITIONING

Basic Concept

A third major behavioral counseling approach is based on the concepts of classical conditioning. *Classical conditioning* is the name given to those behavioral intervention methods based on the idea that the events that *precede,* or come before, the targeted behavior are the things that control the behavior. These events can be represented symbolically by the first part of the behavioral formula ($S^D \rightarrow R \rightarrow S^R$). These techniques are frequently labeled stimulus-response, respondent, or type-S conditioning methods. The basic law that summarizes stimulus-response methods says that "behavior is controlled by its antecedent stimuli." Because certain stimuli cause certain responses, new stimuli can be associated with the old stimuli that give the desired response, and soon the associated stimuli bring about the given response.

Classical conditioning methods are based on the work of a Russian physiologist, Ivan Pavlov (1927), who, while studying the salivation process of dogs, noticed that food brought about the salivation response in the animals. When he made a certain noise at the same time he introduced the food, he discovered that, after repeated pairing, the noise became capable of producing the saliva.

Four elements must be present for classical conditioning to occur: first, an unconditioned event or stimulus (UCS), such as the food in Pavlov's experiments; second, the unconditioned response (UCR), which follows the unconditioned stimulus, or the saliva in Pavlov's studies; third, a conditioned stimulus (CS), which is a natural event, such as Pavlov's noise, that is introduced at the same time as the unconditioned stimulus (UCS); and fourth, a conditioned response (CR), which has been matched or associated in some way with the unconditioned response (UCR).

This law can be illustrated as follows:

$$
\begin{aligned}
\text{UCS} &\rightarrow \text{UCR} \\
\text{(CS) (UCS)} &\rightarrow \text{(UCR)} \\
\text{(CS)} &\rightarrow \text{(UCR) (CR)} \\
\text{CS} &\rightarrow \text{CR}
\end{aligned}
$$

The repeated pairing of the conditioned stimulus (CS) with the unconditioned stimulus (UCS) ultimately causes the conditioned stimulus (CS) to bring about the unconditioned response (UCR). It is also possible to match the unconditioned response (UCR) with a conditioned response (CR) so that ultimately the conditioned stimulus (CS) can bring about a conditioned response.

Stimulus-response conditioning is a very important learning tool and has affected us in a myriad of ways. For example, if every time you were awakened by an alarm, someone in your house was simultaneously cooking bacon, you would soon wake up with the desire for bacon even when it was not being prepared. Another example is that of a young child who was bitten by her cousin's dog. At the sight of that dog or one similar to it, she would probably cry or show another type of anxious or tense behavior.

Stimulus Conditioning Phenomena

Several major learning processes can occur as a result of stimulus control methods. They are stimulus generalization, stimulus discrimination, reinforcement, extinction, and spontaneous recovery.

Stimulus generalization takes place when a response that has been associated with a particular stimulus occurs after other similar stimuli cues are presented. In other words, an association of similar discriminative stimuli has occurred, and both the original stimuli and similar stimuli elicit the same response; for example, a student who has done poorly in a particular math class may generalize and want to avoid all math classes in the future.

Stimulus discrimination occurs when similar discriminative stimuli do not bring about the same response. In other words, a discrimination has occurred between and among the various stimuli; for example, a student who has done poorly in a particular math class may want to avoid the teacher who taught that math class but does not want to avoid other math classes or teachers.

Reinforcement occurs when the matching of the conditioned and unconditioned stimulus is repeated. In other words, the association between the unconditioned stimulus and the conditioned stimulus is strengthened; for example, the periodic matching of the alarm and the cooking of bacon is a reinforcement of this respondent conditioning. Note how this term has different meanings in operant and in classical conditioning.

Extinction occurs when the response is no longer elicited as a result of the conditioned stimulus. In other words, the conditioned stimulus is no longer effective in bringing about the desired response—for example, when after a while the alarm does not bring about a desire for bacon. Again, note how this term has distinctive meanings in operant and classical conditioning.

Spontaneous recovery occurs when the conditioned response returns unexpectedly after it has become extinct. In other words, after the set or relationship between the conditioned stimulus and conditioned response appears to have been broken, it reappears; for example, after extinction has occurred, one wakes up to the alarm without the desire for bacon, but spontaneous recovery is evident when suddenly one morning the alarm again brings a desire for bacon.

Uses of Stimulus-Response Conditioning

Conditioning based on the first part of the behavioral formula ($S^D \rightarrow R$) is useful under two different sets of circumstances. First, it is used to modify or change the stimulus that causes a certain behavior. Second, it is used when a relatively neutral stimulus causes an unwarranted response and modification of the response is necessary.

STIMULUS CHANGE. There are two stimulus change methods: discrimination training and stimulus control techniques. Discrimination training is used when it is desirous to have only a very specific behavior cause a specific response. It is accomplished by reinforcing a particular behavior regularly in the presence of a specific stimulus and not reinforcing or extinguishing that behavior in the presence of another specific stimulus; for example, discrimination training is used to teach individuals the fact that red lights signal danger and green lights signal safety.

Stimulus control techniques are used when it is desirous for the response to be eliminated. It is done by removing the discriminative stimulus or by developing a new pairing pattern so that the unwanted behavior occurs only in the presence of an undesired stimulus; for example, stimulus control methods are used when individuals avoid smoking by not buying cigarettes or by not associating with friends who smoke.

CHANGING RESPONSES.   When relatively neutral stimuli bring about an inappropriate or undesired response, the response pattern must be changed. This type of respondent conditioning is often used with clients who have a high degree of anxiety, fear, phobia, or avoidance. The object is to replace the conditioned response (tension or anxiety) rather than the stimulus. This conditioning relies on the fact that the organism cannot respond in incompatible ways at the same time (tension and relaxation). Therefore, it is possible to desensitize or inhibit the learned response (tension) by substituting an incompatible newly learned response (relaxation). There are four major procedures used in changing response patterns. They are systematic desensitization, assertiveness training, flooding and implosive therapy, and aversion methods. Another method that appears to be based on similar principles but is usually not included in respondent conditioning is paradoxical intention.

SYSTEMATIC DESENSITIZATION.   Systematic desensitization can be described as a counterconditioning or deconditioning process that is used to eliminate negative feelings, anxiety, or other aversive reactions to various stimuli that elicit these unpleasant emotions. Systematic desensitization has been used extensively by Joseph Wolpe (1958, 1982), who employed this technique with phobic clients. It is based on two concepts. The first is that two opposite emotive states, anxiety and relaxation, cannot exist at the same time. Second, if anxiety can be eliminated or inhibited in the presence of those situations that produce it, those situations will eventually lose their power to evoke anxiety in the future. This latter notion is referred to as the *reciprocal inhibition principle*. It is used with individuals whose anxiety is so intense that they are incapable of doing the things they want to do or need to do; for example, the student whose fear of dogs prevents her from visiting friends who own a dog, or the person with a disability whose fear of trains prohibits him from attending a rehabilitation program. The systematic desensitization procedure can be employed to help the student visit her friends or to help the client with a disability attend the rehabilitation program.

Systematic desensitization has been found to be effective with a wide variety of anxiety-related problems in human beings. Among these are fear of heights, open spaces, public speaking, flying, and test taking; school phobia; stuttering; interpersonal anxiety arising from poor interpersonal relationships, such as jealousy and fear of criticism, disapproval, or rejection; psychophysiological illness, such as asthma, headaches, insomnia, speech disorders, and various kinds of sexual dysfunctioning (Kalish, 1981; Schwartz, 1982). The procedure has been effective with both individual and group counseling.

There are three major aspects to this strategy: training clients in deep-muscle relaxation; constructing an appropriate hierarchy of anxiety-provoking situations; and counterpoising relaxation and the anxiety-provoking stimuli. It often takes 6 months or longer to guide a client through this intervention strategy. The typical steps in this procedure are as follows:

1. Identify the target behavior. A comprehensive history of the client should be obtained to analyze the problem and the events related to it. This is done by extensive questioning and through the use of the Life History Questionnaire, which was developed for this purpose (Wolpe & Lazarus, 1966).

2. Determine the factors related to the client's phobic condition. Obtain a detailed description of all the situations related to the fear and discomfort that the client experiences. This is frequently done through verbal discussion; however, the Fear Survey Schedule, the Willoughby Questionnaire, or the Bernsenter Self-Sufficiency Inventory (Wolpe, 1982) may be used.

3. Help the client construct an anxiety hierarchy. Assist the client in identifying and then ranking a list of anxiety-provoking situations. This is done by determining which stimuli cause the least discomfort, then identifying those that cause increasing feelings of discomfort to isolate the situations that cause the most anxiety. These hierarchies should be as real and as concrete as possible and deal with relevant situations, people, times, and places.

4. Teach the client progressive relaxation techniques. This consists of teaching the client how to relax different muscles of the body. In a soothing voice, guide the client through an orderly sequence of relaxing different muscles until a state of complete relaxation is achieved. Clients often need considerable practice before they can relax completely. Plan to give relaxation exercises for homework.

5. Develop a plan for presenting scenes from the anxiety hierarchy to the client. The plan should present different scenarios for each item on the anxiety hierarchy and should have a logical progression from covert scenes having little intensity to overt and more realistic situations. Expect to progress slowly and deliberately, and include homework assignments.

6. Desensitize systematically by presenting the anxiety hierarchy while maintaining relaxation. Progress slowly through scenes from the anxiety hierarchy until the client signals that discomfort is being experienced. At this point, reinforce the relaxation until the client again experiences complete relaxation. The procedure is repeated over and over until the connection between these stimuli and the response of anxiety is eliminated and the fear is neutralized. The process is continued until the situation that originally provoked the anxiety no longer does so.

7. Plan to follow up treatment periodically. Expect to reinforce the treatment when necessary.

There are several variations to Wolpe's technique including in vivo desensitization, which uses real-life situations; covert desensitization, which employs the imagination; contact desensitization, or participant modeling desensitization, which uses observation and simulated experiences; automated desensitization, which employs tape-recorded sessions; and emotive imagery, which uses the imagination to desensitize fears caused by thoughts, images, and wish fulfillments in children.

ASSERTIVENESS TRAINING. Assertive individuals are those who act in their own best interest without too much anxiety and without infringing on the rights of others. Assertive people are aware of their rights; communicate their opinions, needs, and feelings in appropriate ways; and make reasonable demands on others. This behavior has been learned by the assertive person over the course of his or her lifetime. Unassertive individuals allow themselves to be treated as persons of little or no consequence. Typically, unassertive people are taken advantage of by others; lack spontaneity; have difficulty expressing their thoughts, opinions, and emotions; fail to rise to meet unjust treatment; allow others to make decisions in social and work situations; and lack self-esteem (Shaw, Wallace, & LaBella, 1980).

Assertiveness training is a process designed to help clients explore new alternatives that can be open to them. It encourages them to respect their own feelings while, at the same time, respecting the feelings of others. The technique is designed to reduce the fear or anxiety response that has been caused by, or is related to, a particular situation or event and therefore frees the person to express his or her feelings and ideas. Thus, the individual should be able to make more appropriate choices and act more responsibly in his or her daily life. The basic concept involves a counterconditioning process that replaces the fear and anxiety response (the conditioned response) that occurs as a reaction to a particular stimulus or set of stimuli with a new conditioned response (assertive behavior). The principal method used in assertiveness training is counterconditioning; however, other effective intervention strategies such as modeling, role-playing, role reversal, direct instruction, coaching, and contingency reinforcement are often used along with this deconditioning process.

Learning to become assertive is hard work for some clients and extremely difficult for others. Those who encounter the most difficulty are often enmeshed in an environment that reinforces unassertive behavior and punishes assertive behavior. In these cases, clients will have to work on modifying the environment as well as their own behavior. Sometimes clients confuse assertiveness with aggression. They need to learn that aggression is quite the opposite of assertiveness and that aggression is really violating the rights of others by strongly acting out one's desires or frustrations in a hostile manner.

The usual steps employed in this process are as follows:

1. Help the client recognize that his or her inhibitions are causing a great deal of tension and unpleasantness. The client must learn to overcome these inhibitions while respecting the rights of others.

2. Obtain detailed descriptions of all the situations that are related to unassertive behavior, and identify specific instances of unassertiveness and the stimuli that cause and reinforce the behavior.

3. Help the client arrange a hierarchy, from the situations where the client has a higher probability of being assertive to those where the client is non-assertive.

4. Teach the client the distinction between assertiveness and aggression, and between unassertiveness and politeness. Help the client identify and accept his or her own personal rights as well as the rights of others by helping the client clarify his or her understanding of what is appropriate behavior in different situations.

5. Develop a plan to teach the client more assertive behavior. Use covert and symbolic means with weak stimuli or mundane events initially, and then have the client practice assertive behavior. Plan to move on to more intense stimuli and more overt situations. In these graduated situations, incorporate modeling, role-playing, role reversal, direct instruction, and contingent reinforcement.

6. Implement the plan. Be systematic and provide positive feedback and reinforcement. Do additional role-playing, modeling, homework, and practice sessions as needed. Do not rush the client. Instruct the client to do homework and take notes on his or her behavior and report back on how he or she felt and how the assignment went. Provide assertiveness training for a variety of situations; otherwise generalization is less likely to occur.

7. Encourage the client to evaluate his or her own behavior and any changes that have taken place. Plan to follow up with the client periodically. Reinforce the treatment when necessary.

FLOODING. The term *flooding* of and by itself indicates that too much of something is present. In counseling it involves exposing a client to a stimulus, then repeatedly or gradually increasing the time or the intensity of the experience without allowing the client to escape or avoid the exposure. It can be used to extinguish or lessen certain behaviors, or it can be used to help clients overcome behavioral deficits. When it is used to overcome a behavioral deficit, the assumption is made that the client has had a prior real or imagined experience that paired the target behavior with an aversive consequence. Hence, the client will want to avoid or escape from these situations or things that evoked the anxiety or fear. The process involves repeated exposure to the feared stimulus until the client learns that no aversive consequences will follow (Groden & Cautela, 1981). For example, a person who cannot ice skate because of fear of having an accident can be taught to overcome this fear through the covert method of imagining himself or herself skating and watching himself or herself fall on the ice over and over and over again.

When the method is used to help with a behavioral excess, the assumption is made that certain conditions will foster or cause inappropriate responses. The process is designed to present the inappropriate behavior for long periods or in massive doses so that the individual will get exhausted or tired of the behavior (Rachman, Hodgson, & Marks, 1972). This exhaustion will teach the client to act differently when presented with the same or similar conditions; for example, a student who constantly taps a pencil or chews gum can be taught to avoid this behavior by being kept after school and instructed to tap the pencil or chew gum continuously for prolonged periods. The technique, sometimes referred to as the *massing of trials,* employs the following steps:

1. Identify the target behavior by obtaining an extensive client history.
2. Obtain detailed descriptions of all the situations that are related to the maladaptive behavior.
3. Ask the client to arrange in hierarchical order scenes that facilitate the maladaptive behavior from most likely to least likely.
4. Plan a flooding program. From the client's history and situational hierarchy, plan an intervention at a scene that is most likely to cause the maladaptive behavior. Plan how to use covert, symbolic, and overt means to present the stimuli in massive doses until avoidance is eliminated or anxiety reduced. Plan homework assignments.
5. Activate the plan. If appropriate, combine with fading to remove other prompting cues. Provide positive feedback as a systematic contingent reinforcement. Do not allow the client to escape from the stimuli.
6. Plan to follow up with the client periodically. Reinforce when necessary.

Flooding, as a technique, has other side effects; for example, using this strategy with the behavior of excessive cigarette smoking does produce the exhaustion phenomenon, but it also causes more tar and nicotine to be inhaled, which is poor for the health of the client. Consequently, this technique has to be carefully thought out prior to its use to minimize or avoid the side effects. A form of flooding that relies on cues and covert procedures that are frightening to the client is called *implosive therapy* (Shipley, 1979).

AVERSION TECHNIQUES. Aversion techniques can be employed when the counselor wants to assist the client in eliminating or stopping an undesired behavior. The undesirable behaviors may be behavioral excesses, such as overindulgence in food, sweets, or alcohol, or they may be unwanted behaviors, such as compulsive gambling, self-injurious body behavior, cigarette smoking, or enuresis. The basic concept in aversion methods is repeatedly pairing an aversive or noxious stimulus with the undesired behavior. This association of the painful or noxious stimulus with the unwanted behavior should ultimately lead to the cessation of the target behavior. For example, a person who smokes excessively is taught to snap a rubber band against his or her wrist when the urge to smoke occurs. This repeated pairing of the pain and the urge to smoke is designed so that ultimately the urge to smoke will elicit pain and thus curtail the smoking habit. Aversive techniques have been used throughout human history to bring about desired changes (Kazdin, 1978). The ancient Greeks and Romans used such techniques to cure tics and excessive alcoholism. These methods have proved to be successful with many behavioral excesses such as drinking, eating, and smoking. These behaviors are paired with aversive stimuli such as electric shock or a nauseating incident.

Aversive methods should be well thought out before their use for several reasons. First, care should be given to minimize the possibility that the client will use avoidance, escape, or aggression in reaction to the stimulus. Second, every effort should be made to keep the strength or the intensity of the stimulus as low as possible because it can be the actual presence of the aversive or noxious stimulus that brings about the desired change and not the strength of the stimulus. Third, the counselor should try to ensure that the stimulus is truly aversive to the client and not neutral or positive. And fourth, repetition and follow-up need to be incorporated into the treatment so that the aversive stimulus does not have transient or limited-time effects. As a general rule, aversive techniques should be considered only when an individual's biological or psychological well-being is being severely affected by the inappropriate behavior and other treatment alternatives are not appropriate.

Typically, an aversion intervention program uses the following steps:

1. Identify the target behavior by obtaining a detailed client history.
2. Have the client describe the maladaptive behavior and the situation that is normally related to the behavior.
3. Plan what aversive stimulus should match with the undesired behavior. What have other investigators used in similar circumstances? Does the aversive stimulus have any negative side effects? Plan how the pairing will occur, how much homework to assign, and how records will be kept. Plan to use covert, symbolic, and overt techniques, and start with less intensive stimuli.
4. Implement the program. Associate the unpleasant stimulus as closely with the inappropriate behavior as possible. Make sure that the induced stimulus is presented when and only when the unwanted behavior occurs to ensure a strong association.
5. Monitor the program continually. Individuals differ in their responses to aversive stimuli, so careful supervision is required.
6. Continue the conditioning until the inappropriate response no longer occurs in an overt or natural setting or in circumstances where the behavior previously occurred most frequently.
7. Plan to follow up with the client periodically, and reinforce the pairing when necessary.

The aversion technique is often best combined with other contingency management techniques that are designed to initiate or increase the likelihood of another behavior. Thus, although one behavior is being curtailed, another is taking its place; for example, by this process clients can be taught to eliminate fattening foods and eat more nutritious ones.

PARADOXICAL INTENTION. In this technique the counselor instructs the client to engage in a behavior that appears to be incompatible with or the direct opposite of the desired goal. This strategy is called *paradoxical* because it is designed to eliminate a problem behavior by the unusual means of encouraging it. The technique has been used extensively by Frankl (1960, 1985), who employed it with clients who manifested phobias, obsessions, and anticipatory anxiety. Individuals who manifest a fearful expectation of an aversive reaction become enmeshed in a vicious cycle. The anxiety causes the aversive reaction to occur, and thus the thing that the client fears does in fact happen. In dealing with clients who became immobilized, Frankl required them to intend to act toward the item or event that would raise the unwarranted fear. This would be done in a humorous way. He discovered that this experience, which he reinforced in a variety of ways, brought about a change in attitude toward the item that caused the anxiety, and the symptom, or that response, would be diminished.

Paradoxical intention is designed to change the client's attitude toward his or her behavior; thus, this method can and has been used for a variety of problems. Some have used it to reduce stress (Shoham-Salomon & Jancourt, 1985), treat insomnia (Espie & Lindsay, 1985), or work with alcoholic families (Held & Heller, 1982). Others (Dreikurs, 1967; Grunwald & McAbee, 1985; Sweeney, 1989) have used this approach in working from an Adlerian frame of reference and have illustrated its use with clients whose maladaptive behavior is an attention-getting device. For example, a client who screams with a loud voice at inappropriate times is told by the counselor to scream as loud as he or she can. As the client responds, the counselor teases the client by saying something like, "Oh, that only earned you a 'C.' I know you have an 'A' scream inside of you." This is periodically repeated until the inappropriate behavior stops. This technique is often used in family counseling, in which the counselor instructs the parents on how to use the technique. The usual procedures involved are:

1. Identify the inappropriate behavior.
2. Persuade the client to produce the behavior in the most intense way possible.
3. Inject humor into the situation as the client engages in the behavior. This allows the client to become somewhat detached from the problem by laughing at it.
4. Repeat Steps 2 and 3 until the inappropriate behavior is minimized.

As with other techniques, paradoxical intention is not a panacea. It does not always work, nor can or should it always be used. Its effectiveness can be enhanced by using it with other procedures to teach clients more appropriate ways of behaving.

|| SUMMARY

The essential concepts of behaviorally focused interventions were briefly outlined in this chapter. Behavioral techniques are based on the premise that all behavior is learned and individuals develop habits and reinforce their behavioral activities. These strategies are designed to help clients learn a new behavior, increase the likelihood of a present behavior, or eliminate an unwanted or undesirable behavior. Behavioral counselors focus on concrete behaviors and activities, and they employ very specific, highly structured, goal-directed, and sequentially ordered steps to assist clients. Homework assignments are often given. Contingency contracts, which outline the goals, responsibilities, and contingencies involved in the counseling process, are frequently employed. Behavioral approaches include the concepts of observational and simulated learning, which promote the acquisition of new skills by using modeling and role-playing strategies; the operant conditioning principles, which explain the modification of behavior by the appropriate use of reinforcement and punishment contingencies; and the classical conditioning theory, which gives the basis for strategies such as stimulus control, systematic desensitization, assertiveness training, flooding, aversion techniques, and paradoxical intention.

After reading this chapter on the behaviorally focused strategies, you should be able to identify the major aspects of the social learning, operant conditioning, and classical conditioning interventions. You should be able to understand how and when to use these strategies, discuss the sequential steps involved in applying a behavioral technique, and describe the contents of a contingency contract. In order to accomplish a high level of facility in using these methods, you will have to study these techniques further and obtain supervised practice in applying these strategies with actual clients.

REFERENCES

Bandura, A., & Walters, R. H. (1963). *Social learning and personality development.* New York: Holt, Rinehart & Winston

Dreikurs, R. (1967). *Psychology in the classroom.* New York: Harper & Row.

Espie, C., & Lindsay, W. (1985). Paradoxical intention in the treatment of insomnia. *Behavior Research Therapy, 23,* 703–709.

Frankl, V. E. (1960). Paradoxical intentions: A logotherapeutic technique. *American Journal of Psychotherapy, 14,* 520–525.

Frankl, V. E. (1985). Paradoxical intention. In G. R. Weeks (Ed.), *Promoting change through paradoxical therapy* (pp. 99–110). Homewood, IL: Dow Jones-Irwin.

Groden, G., & Cautela, J. R. (1981). Behavior therapy: A survey of procedures for counselors. *The Personnel and Guidance Journal, 60,* 175–180.

Grunwald, B. B., & McAbee, H. V. (1985). *Guiding the family.* Muncie, IN: Accelerated Development.

Held, B., & Heller, L. (1982). Symptom prescription as a metaphor: A systematic approach to the psychosomatic-alcoholic family. *Family Therapy, 9,* 133–145.

Kalish, H. I. (1981). *From behavioral science to behavioral modification.* New York: McGraw-Hill.

Kanfer, F. H., & Goldstein, A. P. (1986). *Helping people change: A textbook of methods* (3rd ed.). Elmsford, NY: Pergamon Press.

Kazdin, A. E. (1978). *History of behavior modification.* Baltimore, MD: University Park Press.

Kazdin, A. E. (1989). *Behavior modification in applied settings* (4th ed.). Pacific Grove, CA: Brooks/Cole.

Krumboltz, J. D., & Thoresen, C. E. (1976). *Counseling methods.* New York: Holt, Rinehart & Winston.

Martin, G., & Pear, J. (1983). *Behavior modification: What it is and how to do it* (2nd ed.). Englewood Cliffs, NJ: Prentice-Hall.

Pavlov, I. P. (1927). *Conditioned reflexes: An investigation of the physiological activity of the cerebral cortex.* (G. V. Anrep, Trans.) London: Oxford University Press.

Premack, D. (1965). Reinforcement theory. In D. Levine (Eds.), *Nebraska symposium on motivation* (pp. 123–180). Lincoln, NE: Nebraska Press.

Rachman, S., Hodgson, R., & Marks, I. M. (1972). The treatment of chronic obsessive-compulsive neurosis: Follow-up and findings. *Behavior Research and Therapy, 10,* 181–189.

Schwartz, A. (1982). *The behavior therapies.* New York: Free Press.

Shaw, M. E., Wallace, E., & LaBella, F. (1980). *Making it assertively.* Englewood Cliffs, NJ: Prentice-Hall.

Shipley, R. H. (1979). Implosive therapy: The technique. *Psychotherapy: Theory, Research, and Practice, 16,* 140–147.

Shoham-Salomon, V., & Jancourt, A. (1985). Differential effectiveness of paradoxical intention for more versus less stress-prone individuals. *Journal of Counseling Psychology, 32,* 449–543.

Skinner, B. F. (1938). *The behavior of organisms: An experimental analysis.* New York: Appleton-Century-Crofts.

Skinner, B. F. (1953). *Science and human behavior.* New York: Free Press.

Skinner, B. F. (1989). *Recent issues in analysis of behavior.* Columbus, OH: Merrill.

Sweeney, T. J. (1989). *Adlerian counseling: A practical approach for a new decade* (3rd ed.). Muncie, IN: Accelerated Development.

Wolpe, J. (1958). *Psychotherapy by reciprocal inhibition.* Stanford, CA: Stanford University Press.

Wolpe, J. (1982). *The practice of behavior therapy* (3rd ed.). Elmsford, NY: Pergamon Press.

Wolpe, J., & Lazarus, A. (1966). *A behavior therapy technique: A guide to the treatment of neuroses.* Elmsford, NY: Pergamon Press.

PART 3

The Counselor's Role Communication Skills

Counselors-in-training develop their counseling skills through a process that involves both didactic and experiential learning. The didactic process focuses on cognitive learnings in which students read the technical literature, hear about various theoretical counseling approaches, see demonstrations of these approaches, and study typescript, audiotaped, and videotaped vignettes of actual counseling case material. The experiential process focuses on learnings obained from direct experience in which counselors-in-training develop their skills through modeling the behavior of admired experts; practicing appropriate counselor response patterns; obtaining peer feedback on their counseling practices; and receiving professional supervision of their counseling practices in the typical sequential learning steps of role-playing, practicum, and internship.

Typically, students entering their first experientially based course in counseling are confused about how to begin a counseling relationship. They may be familiar with the theoretical literature and may be able to discuss intellectually Adlerian, behavioral, Rogerian, and some other major approaches to the counseling process. Nevertheless, they need to develop important communication skills to be effective counselors.

Counseling involves a dynamic communication process between two people who are interacting with one another. This interactive process is a collaborative effort in which you and the client undertake certain roles, responsibilities, and behaviors. In this collaboration, you as the counselor must take the responsibility for providing a facilitative climate and enhancing the client's motivation to change. Thus, the burden is on you to employ appropriate verbal, paraverbal, and nonverbal communication skills to influence the direction, the duration, and the eventual effectiveness of the counseling process. The client must feel free enough to reveal his or her real concerns and eventually must learn that he or she is the only one who can assume the ultimate responsibility for bringing about the desired changes.

As the counselor, you must be concerned about both perceiving what the client is attempting to communicate and responding to that message in an appropriate manner. Client messages have cognitive as well as affective components.

The cognitive content of a message is normally fairly easy to identify. On the other hand, clients frequently communicate vague and incomplete pictures of themselves and their concerns. They may exhibit a lack of congruity between their verbal and nonverbal signals or a tendency to verbalize about irrelevant material. Clients also frequently show their lack of personal ownership of a problem by speaking about an unidentified person or group (for example, "*It* is a common fault . . ."; "*Some people* believe that . . ."; or "*Everybody* cannot stand for . . ."); or by speaking for someone who is not present (for example, "*Bill* believes . . ."; *John* says . . ."; *Mary* does . . ."). Speaking for others may or may not represent the other person's position, but it can distract from the major focus of the conversation. While listening to the cognitive component of the client's message, you will need to be aware of these tendencies to avoid or deflect the problem.

The affective component of a client's message is the manifestation of the internal reaction that individuals have to their experiences. Clients may identify their feelings by labeling them or describing how they feel, or they may vaguely allude to how they feel. Frequently, these internal reactions are communicated by outward signs, such as crying when one is sad or extremely disappointed, shouting when one is angry, and smiling when one is happy or has experienced a pleasant event. Some clients have difficulty expressing themselves, particularly when they have a fear of being rejected or not being taken seriously. Feelings are more frequently communicated through nonverbal and paraverbal means than they are by verbal channels. You will need to pay attention to your client's facial expressions, voice inflections, and other cues. You should also be aware of the vagueness that may be present in some messages and the contradictions that can exist between the verbal and nonverbal channels. You must make every effort to recognize and accept your client's feelings in order to facilitate the client's awareness and expression of these inner reactions, any conflicts that he or she may experience about these feelings, and any barriers that may prevent the expression of these feelings.

Barriers to effective dialogue between counselors and their clients can arise from causes such as making value judgments about clients, having selective or erroneous perceptions of the issues that clients present, and giving inappropriate responses to client messages. You should avoid making value judgments about your clients, labeling them or their actions, or questioning their motives. When issues arise that require evaluations of your client's thoughts, feelings, or behaviors, or an analysis of your client's motives, you should focus on helping your client make these judgments for himself or herself so that effective growth can occur. Because it is often difficult to really feel the problems and sense the difficulties that another person is experiencing, selective or erroneous perceptions can occur. Your experiences of life are different from your client's. Hence, as a counselor you must try to understand the perceptual views of your clients who, because of different cultural backgrounds, personal interests, age, and other factors, may have different perceptions of their experiences and quite distinctive meanings for certain words and phrases. Learning to respond to clients in appropriate ways is a skill that can be learned. This section of the text is devoted to helping you as a counselor-in-training understand, develop, and master the important communication skills used in counseling.

COMMUNICATION AS A ROLE FUNCTION

The counseling relationship is based on effective communication between two people. The client is the person who seeks some resolution of a problem. The counselor is the skilled professional who uses his or her skills and knowledge of human behavior to assist the person in need. Because the purpose of this interaction is to improve the client's well-being, the burden is on you as the counselor to employ appropriate communication skills that will influence the shape, the duration, and indeed the eventual effectiveness of the counseling process.

Considerable attention has been focused on improving the communication skills of counselors since Robinson (1950) outlined a verbal response system based on the degree of lead involved. His use of the term *lead* implies that counselors have the responsibility of employing responses that anticipate the client's needs and enable the client to progress further in the interviewing or counseling process. These efforts to develop a skills-oriented approach to the education and training of counselors have led to:

1. the identification of a number of distinctive types of verbal responses that counselors employ within the counseling dyad (Benjamin, 1987; Brammer, 1988; Danish, D'Augelli, Hauer, & Conter, 1980; Hackney & Cormier, 1988; Hill, 1978; Hoffman, 1959; Spooner & Stone, 1977);
2. the development of several scales to evaluate counselors' skills in using responses (Danskin, 1955; Hill, 1978; Hoffman, 1959; Spooner & Stone, 1977); and
3. the formulation of several programs that systematically teach these skills (Carkhuff, 1969; Danish et al., 1980; Ivey & Authier, 1978; Kagan, 1972a).

Counselor education programs that have used the skill-oriented approach typically have their students learn a repertoire of helping in prepracticum courses. These skills are further developed in later courses until, as Kagan (1972b, p. 44) pointed out, a critical phenomenological change occurs when the trainee "puts it all together" and "becomes truly capable of therapeutic intervention."

These efforts to develop counselor skills have been found to be useful for understanding the counseling process and have been effective in counselor training programs (Carkhuff, 1969; Danish, D'Augelli, & Brock, 1976; Hill, 1978). The available programs have emphasized the basic skills that are important in the initial phases of counseling (Hill, 1978). These systems need to be developed and expanded to provide adequate methodologies for the counselor's paraverbal and nonverbal, as well as verbal, modalities, and to provide a comprehensive system for describing skills throughout the entire counseling process (Blum & Rosenberg, 1968; Hill, Charles, & Reed, 1981; Lambert, DeJulio, & Stein, 1978).

This section of the text presents a two-dimensional conceptual model called the *role communication skill model*. Because communication presupposes role-taking opportunities (Kohlberg, 1969) and because counseling is a dynamic interactive communication process, the first dimension used to describe the counselor's functional behaviors is the *role* construct. This construct was employed earlier by Danskin (1955), Hoffman (1959), Muthard (1953), and Robinson (1955) in their investigations of the types of verbal responses used by counselors. In this interactive communication process, it is the counselor who must take the responsibility for acting or behaving in ways that will provide the facilitative climate and the conditions necessary to enhance the probabilities for client change. Furthermore, because counselors engage in different behaviors and act in several ways in the various stages of counseling, these distinctive ways of behaving are considered to be unique roles. The major counselor roles have been described as:

- attender,
- clarifier,
- supporter or reassurer,
- informer or describer,
- prober or inquirer,
- motivator or prescriber,
- evaluator or analyzer, and
- problem solver (Doyle, 1982).

In this model, roles are seen as the critical or most important behavioral messages, and verbal responses or counselor leads, such as those mentioned by Buchheimer and Balough (1961), Hill (1978) and Spooner and Stone (1977), are seen as having secondary importance. Thus, verbal responses are subsidiary skills necessary for each role. The model includes the four response modes (paraphrasing, approval, self-disclosure, and interpretation) that Hill and associates (1988) found to be the most helpful. And it incorporates the six responses that Elliott and associates (1987) reported as being critical because there were significant differences in the use of these responses among seven diverse therapeutic approaches. These responses were: questioning, providing information, advising, using reflections, interpretations, and self-disclosures.

Furthermore, because communication takes place through paraverbal and nonverbal channels as well as through words, a particular verbal response can convey a different role depending on the counselor's attitude, expertise, and intention. This variation in meaning of a particular phrase appears to be in agreement with Greenberg's (1986) contention that speech acts have different meanings depending on the context or the episode in which they occur. Paraverbal elements include tonal quality, such as intensity, pitch, amplitude, velocity, and raspingness, and voice differentiators, such as laughter, sobbing, and a cracking or breaking voice. Nonverbal messages are communicated by various facial and other body movements. Messages are sent by such things as eye contact, nodding the head, and manipulating the facial muscles to produce frowns, indifference, quizzical looks, and smiles. Other nonverbal messages are sent by posture, muscle tone, twitching, and gestures. Changes in the meaning of a particular verbal response and hence the role used by the counselor are thus affected by tonal quality as well as gestures, body movements, and facial expressions. The eight counselor roles and the subsidiary verbal responses associated with these eight roles are outlined in further detail in the chapters of this section.

The second dimension employed to describe counselor behaviors is the *level* construct. This dimension is employed as a method for the qualitative evaluation of the counselor's competence and timing in using a particular role with appropriate nonverbal and paraverbal characteristics. Qualitative phenomena, such as therapeutic knowledge, insight, accuracy, and appropriateness in terms of pace and timing in using the role, can be accounted for with this construct. Furthermore, the level construct provides a methodology for indicating whether the paraverbal and nonverbal channels of communication are used appropriately.

The counselor's effectiveness can be ascertained by employing a competency continuum with a Likert-type scale. For convenience, a four-point scale ranging from 1 to 4 is used. The four points are:

Level 1: Poor use of the role,
Level 2: Mediocre use of the role,
Level 3: Good use of the role, and
Level 4: Excellent use of the role.

The overall Level-of-Response Scale for rating the counselor's proficiency in using appropriate responses is indicated at the end of this section. The specific competency criteria for rating the counselor-in-training's use of each role is incorporated in the chapters that follow. Counselors-in-training sometimes prefer to use the descriptive terminology—poor, mediocre, good, or excellent—rather than the numerical rating.

The level construct was first proposed by Carkhuff (1969), who outlined a five-point scale for rating the counselor's effectiveness in both discriminating and communicating responses. Responses rated below 3 were considered to be distractive and had the potential for doing more harm than good. A response with a rating of 3 was considered to be minimally facilitative, and a response above 3 was thought to be additive—that is, quite helpful to the client and the counseling process. Gazda and associates (1984) revised Carkhuff's scale to a four-point one. The Level-of-Response Scale used in this text is similar to the scales employed by Carkhuff and Gazda in terms of effectiveness, but it is based on the role construct rather than the "facilitative conditions" stressed by these authors.

The role communication skills discussed in the four chapters of this section have been arranged along a continuum having four major domains. The primary role communication skills—attending, clarifying, and supporting—are presented in Chapter 7. These roles focus on accepting and understanding the client and the client's frame of reference, and communicating warmth, interest, and respect to the client. The intermediate role communication skills—providing information, inquiring, and the use of silence—are described in Chapter 8. These roles require the counselor to give more direction to the counseling process by describing phenomena, probing for more information, and using silence as an effective tool. Chapter 9 outlines the advanced role communication skills— motivating and evaluating. These skills require that the counselor use more direct, deliberate, and rather forceful responses. And Chapter 10 discusses the skills that are needed in the problem-solving role, when the counselor uses specific cognitive, affective, or performance intervention strategies. Hansen, Stevic, and Warner (1986), Robinson (1950), and Shertzer and Stone (1980) discuss similar concepts of a continuum of counselor responses using counselor lead as the construct.

Each role is discussed in some detail. First, the purpose of the role and what it is designed to accomplish are described. Second, the importance of the role is indicated and some guidelines in using the role are outlined. Third, the typical verbal responses associated with each role are given, and illustrations of these responses are presented. Fourth, a methodology for evaluating the effectiveness of the use of this role is discussed, and examples of effective and ineffective responses are shown. Finally, a variety of exercises designed to help you as a counselor-in-training develop these skills is provided. Initial practice in using these skills sometimes feels awkward and clumsy, and you may find yourself focusing on your own words and phrases rather than on the needs of the client. Learning these communication skills requires practice, and repeated practice is usually necessary to master these skills.

In learning these role communication skills, you will need to keep in mind several general principles. First, you should avoid allowing any of your personal concerns to creep into the counseling process and distract from the client's concerns. Second, you must help your client keep the focus on himself or herself rather than on a third party who is not present in the session. Third, you should carefully observe whether your client's nonverbal and paraverbal signals are in agreement with his or her verbal statements. Fourth, you should pay careful attention to any emerging themes and repeating thoughts, feelings, or behaviors that may reveal overriding patterns for your client. Fifth, you must be consciously aware of how your own inner perceptions, values, and experiences affect your interpretation of your client's statements. Finally, you should make every effort to use language that is appropriate to the cultural experience and the educational background of your client.

The exercises in each role have been designed to enable you as a counselor-in-training to learn and practice a variety of verbal responses within that role; to develop the skills necessary to discriminate between low- and high-quality responses; and to practice communicating high level responses. Each role contains three distinctive types of practice exercises: the discrimination exercises, the communication exercises, and the solo exercises. The discrimination exercises are designed to help you

- practice identifying a client's initial or stated concern;
- understand that there are different methods of responding to perceived needs with empathy, respect, and warmth, and practice identifying these different types of responses; and

- recognize that responses are truly different in the degree of their effectiveness, and practice discriminating among different levels of responses.

The communication exercises have been planned to provide the opportunity for you to practice giving appropriate high level responses to a variety of client statements. For each communication exercise, you will be asked to give several different types of appropriate verbal responses. The solo exercises are intended to give you further opportunities to practice your skills under supervisory conditions.

The exercises have been used in a number of ways under the supervision of experienced counselors. Some supervisors have assigned the exercises as homework, others have assigned counselors-in-training to triads for practice, whereas others have had small groups discuss the exercises intensively. The exercises have been found to be most useful when the reasons for the ratings are discussed in the group. Obviously, an exercise in print form does not adequately convey any nonverbal aspects, nor does it reflect any tonal quality. These communicative channels can be employed in ways that convey very distinct meanings. It is, therefore, critical that the exercises be used as an *aid* in the training program and not as a substitute for intensive supervision. Indeed, it is imperative that the responses given by the counselors-in-training be discussed fully and completely in a supervised small-group discussion. The nuances and the reasons counselors-in-training give for their responses and the ratings they assign to these responses are the focal point of the exercises.

Level-of-Response Scale

Because there are qualitative as well as modal differences in responding to clients, the counselor's proficiency in using each of the roles can be thought of as lying along a continuum with four integral points: poor (1); mediocre (2); good (3); excellent (4). The following descriptions should be used as a guide in rating the counselor's responses. Note that responses may be rated with a decimal rating (that is, 2.3, 2.5, 2.9) depending on the closeness or distance from the ordinal integers given in the rating scale.

Level 1: Poor Use of the Role. The counselor does not attend to the client's concerns or uses a role that is inappropriate at this point in the counseling process. The reflective, paraphrasing, or clarifying responses have little or no relationship to the content and feeling expressed by the client; the counselor is informing, describing, or explaining things incorrectly; irrelevant material is sought; the probing or motivating roles are used prematurely; or the counselor is using a particular problem-solving technique incorrectly or inappropriately. The counselor may show lack of interest, may ridicule, or may try to punish the client.

Level 2: Mediocre Use of the Role. The counselor partially attends to the feelings or content expressed by the client, or a role is used that is not very helpful at this stage of counseling. The reflective, paraphrasing, or clarifying responses are superficial interpretations of the client's cues. The counselor may provide the client with partial or incomplete information, or probe for appropriate data but not at a sufficient depth. When the motivating role is used, it is done in a weak manner. The problem-solving role is used prematurely, incorrectly, or incompletely. The counselor may show minimal interest, may give cheap advice, or may ask rather meaningless questions.

LEVEL 3: GOOD USE OF THE ROLE.   The response mode attends to the stated content and to the feelings of the client, or the counselor uses a role that is very suitable at this point in the counseling process. The counselor interprets messages correctly and communicates this effectively to the client. The attending and clarifying roles are used appropriately, phenomena are explained correctly, and probes are topical and at sufficient depth. Motivational statements are adroit, and the problem-solving role is used competently. The counselor shows a high degree of interest, makes appropriate inquiries, and interprets the verbal, paraverbal, and nonverbal messages in meaningful ways.

LEVEL 4: EXCELLENT USE OF THE ROLE.   The response mode used goes beyond attending the stated content and the surface feelings of the client, or the counselor uses a role that is very appropriate and advances the counseling. The counselor interprets the client's messages fully and responds adroitly. Factual information and descriptions are given completely and in an understanding manner. The probing and motivational roles are used when called for and facilitate the progress of the counseling. The problem-solving role is used in a timely way, and the counselor demonstrates his or her mastery of the technique employed. The counselor reveals an intensive interest, makes wise inquiries, and takes some appropriate risks in the interpretation of the verbal, paraverbal, and nonverbal messages.

REFERENCES

Benjamin, A. (1987). *The helping interview* (4th ed.). Boston: Houghton Mifflin.

Blum, A. F., & Rosenberg, L. (1968). Some problems involved in professionalizing social interaction: The case of psychotherapeutic training. *Journal of Health and Social Behavior, 9,* 72–85.

Brammer, L. M. (1988). *The helping relationship* (4th ed.). Englewood Cliffs, NJ: Prentice-Hall.

Buchheimer, A., & Balough, S. C. (1961). *The counseling relationship: A casebook.* Chicago: Science Research Associates.

Carkhuff, R. R. (1969). *Human and helping relationships* (2 vols.). New York: Holt, Rinehart & Winston.

Danish, S. L., D'Augelli, A. R., & Brock, G. W. (1976). An evaluation of helping skills training: Effects on helpers' verbal responses. *Journal of Counseling Psychology, 3,* 259–266.

Danish, S. L., D' Augelli, A. R., Hauer, A. L., & Conter, J. J. (1980). *Helping skills: A basic training program* (2nd ed.). New York: Human Sciences Press.

Danskin, D. G. (1955). Roles played by counselors in their interviews. *Journal of Counseling Psychology, 2,* 22–27.

Doyle, R. E. (1982). The counselor's role communication skills, or the roles counselors play: A conceptual model. *Counselor Education and Supervision, 22,* 123–131.

Elliott, R., Hill, C. E., Stiles, W. B., Friedlander, M. L., Mahrer, A. R., & Marigison, F. R. (1987). Primary therapist response modes: Comparison of six rating systems. *Journal of Consulting and Clinical Psychology, 55,* 218–223.

Gazda, G. M., Asbury, F. R., Balzer, F. R., Childers, W. C., Deselle, R. E., & Walters, R. P. (1984). *Human relations development: A manual for educators* (2nd ed.). Boston: Allyn & Bacon.

Greenberg, L. S. (1986). Change process research. *Journal of Consulting and Clinical Psychology, 54,* 4–9.

Hackney, H., & Cormier, L. S. (1988). *Counseling strategies and objectives* (3rd ed.). Englewood Cliffs, NJ: Prentice-Hall.

Hansen, J. C., Stevic, R. R., & Warner, R. W., Jr. (1986). *Counseling theory and process* (4th ed.). Boston: Allyn & Bacon.

Hill, C. E. (1978). Development of a counselor verbal response category system. *Journal of Counseling Psychology, 25,* 461–468.

Hill, C. E., Charles, D., & Reed, K. G. (1981). A longitudinal analysis of changes in counseling skills during doctoral training in counseling psychology. *Journal of Counseling Psychology, 28,* 428–436.

Hill, C. E., Helms, J. E., Tichenor, V., Spiegel, S. B., O'Grady, K. E., & Perry, E. S. (1988). Effects of therapist response modes in brief psychotherapy. *Journal of Counseling Psychology, 35,* 222–233.

Hoffman, A. E. (1959). An analysis of counselor subroles. *Journal of Counseling Psychology, 6,* 61–67.

Ivey, A. E., & Authier, J. (1978). *Microcounseling: Innovations in interviewing, counseling, psychotherapy, and psychoeducation* (2nd ed.). Springfield, IL: Charles C Thomas.

Kagan, N. (1972a). *Influencing human interaction.* Unpublished manuscript, Michigan State University.

Kagan, N. (1972b). Observations and suggestions. *The Counseling Psychologist, 3*(1), 42–45.

Kohlberg, L. (1969). Stage and sequence: The cognitive-development approach to socialization. In D. A. Goslin (Ed.), *Handbook of socialization theory and research* (pp. 347–480). Chicago: Rand McNally.

Lambert, M. V., DeJulio, S. S., & Stein, D. M. (1978). Therapist in interpersonal skills: Process, outcome, methodological considerations and recommendations for future research. *Psychological Bulletin, 85,* 467–489.

Muthard, J. E. (1953). The relative effectiveness of larger units used in interview analysis. *Journal of Consulting Psychology, 18,* 184–188.

Robinson, F. P. (1950). *Principles and procedures in student counseling.* New York: Harper & Bros.

Robinson, F. P. (1955). The dynamics of communication in counseling. *Journal of Counseling Psychology, 2,* 163–169.

Shertzer, B., & Stone, S. C. (1980). *Fundamentals of counseling* (3rd ed.). Boston: Houghton Mifflin.

Spooner, S. E., & Stone, S. C. (1977). Maintenance of specific counseling skills over time. *Journal of Counseling Psychology, 24,* 66–71.

7 | *Primary Role*
Communication Skills

INTRODUCTION

The counselor's primary role communication skills are covered in this chapter. The attending, clarifying, and supporting roles are explained, examples demonstrating the effective use of each of these roles are given, and exercises are provided to allow you to practice and master these skills. After studying this chapter you should be able to:

- Describe the purposes of the attending, clarifying, and supporting roles.
- Discuss the characteristics that distinguish high level responses from low level responses in each of these roles.
- Give examples of the types of responses used in each of these roles and illustrate good and poor uses of these responses in helping interviews.
- Apply these skills in appropriate situations.

THE ATTENDING ROLE

Before you can help another person, you must pay attention and listen to what the other person is saying. *Attending* is the process of trying to understand a client without making any evaluative judgments about the person. It is accomplished when you actively listen to the client in order to discern the client's primary or essential message; demonstrate an interest, acceptance, and respect for the client and the client's internal frame of reference; and communicate by an appropriate phrase or gesture that you understand what the client is saying or is attempting to say. You need to listen to the essential or underlying message expressed by the client and not the superficial words or phrases. This active listening involves focusing on the content of the phrases the client uses, the emotions or affect the client is feeling or expressing, and the body language the client is using, and putting these signals into a composite message in light of the client's background and experience.

Demonstrating interest, acceptance, and respect for the client requires you as the counselor to establish and maintain good eye contact and attentive body posture and to be fully present for the client. Your responses to the client should accurately communicate your understanding and may vary from a nonverbal gesture, such as a nod of the head, to a simple minimal verbal response, such as "mm-hmm" or "I see," to a more verbal response, which reflects the client's feelings and content, to a deeper level paraphrasing response, which goes beyond what the client has stated in an attempt to catch the internal meaning of the client's statement.

Because the attending role involves the process of being receptive and involved with another person, it is the prerequisite for all other responses to clients. It sets the tone for the relationship, demonstrates to the client that the counselor cares and is sincerely interested, and provides the opportunity for the client to discuss whatever is bothering him or her. The way one attends to the client can have a great deal of influence on the client and the counseling process. Good attending responses tend to reduce a client's fear about revealing himself or herself; thus, they decrease the client's defensiveness and increase his or her sense of trust in you and in the counseling process.

Four guiding principles should be followed in communicating an attending response to the client. First, give yourself a chance to reflect before responding; you will need time to integrate the totality of the client's message and to formulate an appropriate response. Second, try to use responses that are brief and to the point rather than long and all-inclusive. Long responses tend to be counterproductive because clients can be distracted by them or give short responses to them. Third, use terms, phrases, and expressions that are familiar to the client. You should employ a vocabulary appropriate to the client's age, educational level, and cultural background. And fourth, be reasonably spontaneous. Although you can take time to think and reflect on the meaning of a client's statement, the silence caused by a pause that is too long can be distracting.

Types of Attending Responses

The following four types of verbal responses are all normally used in this role and are quite helpful in attending to clients.

1. *Simple minimal verbal.* This is a short response, such as "I see," "Uh-huh," or "Mm-hmm," which is considered the verbal equivalent of a head nod (Okun, 1987). It indicates to a client that the counselor is listening and following his or her statements. A simple verbal response can encourage the client to continue talking (Benjamin, 1987) and can have a significant reinforcing value that increases a client's use of a particular word or topic (Hackney & Cormier, 1988).

2. *Reflection of content and feelings.* This is a response that mirrors or reflects to the client the message that the counselor hears. The counselor responds to what is being said, how it is being said, the underlying feelings that

are evident, and the nonverbal body expressions that are communicated. It is important to sense the feelings and attitudes not expressed by the client and to bring these to the surface (Benjamin, 1987). Words identical or similar to the client's are used to convey to the client the essence of the message that the counselor heard.

3. *Accent.* This is a short response that selects part of the client's previous statement and brings it into focus by repeating it. It is said in a tonal quality that will encourage the client to elaborate further on the part of the message that the counselor thinks is most important (Hackney & Cormier, 1988).

4. *Paraphrasing.* This response is a restatement of the client's essential message using different words or phrases that are carefully chosen (Cormier & Cormier, 1991). Because different words are used, this not only communicates the counselor's understanding of the client's message, it also helps clients see their thoughts, feelings, and behaviors from another perspective. In paraphrasing it is helpful to use responses that go beyond the client's verbal message in order to catch the deeper internal meaning. Counselors often use images and analogies to capture this deeper meaning.

Illustrations of Different Types of Attending Responses

The following three examples demonstrate the use of the attending role. Note how the client's statement can be responded to in different ways and still convey warmth and understanding to the client.

CLIENT 1, A 20-YEAR-OLD WOMAN: I feel as if my friends have been letting me down lately. Until recently I was very popular in school—I had a lot of friends, and I enjoyed being with them. But all of a sudden I feel like I've been losing them. I've been snubbed by some, and I really don't know why this is happening.

1. *Simple minimal verbal:* I understand.
2. *Reflection of feeling and content:* You're upset because you seem to have lost some of your popularity.
3. *Accent:* You've been snubbed.
4. *Paraphrasing:* You feel your friends have been ignoring you, and you're confused as to why this is happening. You're really hurt.

CLIENT 2, A TEENAGER: My mother is constantly on my back. Every time I turn around, there's something else she wants me to do or some more advice she wants to hand out. She really burns me up.

1. *Simple minimal verbal:* That hurts.
2. *Reflection of feeling and content:* You're angry at your mother because she wants you to follow her advice and do her errands for her.
3. *Accent:* Your mother burns you up.

4. *Paraphrasing:* You're behind the eight ball, and every time you want to get out from behind, you're shoved back there by your mother.

CLIENT 3, A 23-YEAR-OLD WOMAN: I'm working part time as a dental assistant; it's the only job I could get. It seems that I've spent my whole life wanting and training to be a teacher, but it's impossible to get a job. I hate what I'm doing now, and I'm so afraid I'll be stuck there for the rest of my life and my education will just go down the drain.

1. *Simple minimal verbal:* I see.
2. *Reflection of feeling and content:* You're angry because you have a job you hate and cannot obtain one for which you were trained.
3. *Accent:* You're afraid you will be stuck.
4. *Paraphrasing:* You feel that you are in a rut and the walls are too high for you to get out.

Levels of Attending Responses

The attending role can be used with various degrees of effectiveness. When you use the role effectively, it shows that you understand both the content and feelings expressed by the client, tend to focus on issues and situations that appear most relevant, communicate to the client that you have heard the major message, and often get behind the stated words to capture or sense the client's underlying message. High level responses communicate to the client that you are a person who is sincerely interested in the client and a person to whom the client can relate in a nonthreatening, trustful manner. The role is used ineffectively when the counselor responds only to the surface level concerns expressed by the client. This is often manifested by the counselor's repeating the content of the client's message. This "parroting" usually causes a circular movement in the dialogue and feelings of discomfort on the part of the client (Evans, Hearn, Uhlemann, & Ivey, 1989). Low level responses are often ineffective and may be counterproductive. They are often said in a tonal quality that creates a questioning atmosphere rather than a permissive or encouraging one. Low level responses can also imply disapproval and criticism of the client's thoughts, feelings, or behaviors. On the four-point scale, the use of the role can be rated as follows:

Level 1: Poor use of the role. The counselor is least effective when he or she reveals little or no interest in the client, when messages are glaringly misinterpreted, when responses reveal little or no awareness of the underlying feelings or concerns of the client, when body movements or voice quality is distracting, or when the role is used at an inappropriate place in the relationship.

Level 2: Mediocre use of the role. The counselor's responses are mediocre when he or she attends partially to the feeling or the content expressed by the client; when the response is a superficial interpretation of the client's verbal, paraverbal, and nonverbal messages; when tonal quality or gestures are somewhat misleading; or when the counselor shows only minimal interest in the client.

Level 3: Good use of the role. The counselor is effective when he or she attends to the stated content and feelings, when messages are interpreted correctly and communicated back accurately, and when the counselor manifests a high degree of interest in the client.

Level 4: Excellent use of the role. The counselor is most effective when he or she demonstrates intense interest in the client; interprets messages fully; responds to the underlying feelings and messages expressed by the client; maintains appropriate eye contact, tonal quality, and physical attentiveness; and uses the role at appropriate times.

Illustrations of Different Levels of Attending Responses

There are three client statements below; each statement is followed by four different responses. The responses are rated according to the Level-of-Response Scale, and the reasons for each of these ratings are indicated. Study these examples to see if you agree with the ratings and the reasons stated. You may find that you do not agree with these ratings. Because the nonverbal and paraverbal aspects of the client's statement and the counselor's responses cannot be adequately conveyed by the printed word, it is quite possible that you will read one or more of these statements differently from the person who rated them. Any counselor response may be rated differently depending on how it is heard. It is extremely important that you state the reasons for your rating so that you can discuss any differences with your supervisor and classmates.

CLIENT 1, A HIGH SCHOOL SENIOR: I'm worried about what I'll do in the future, and I was wondering whether you can tell me if there are any good jobs in accounting.

Responses	Rating and Discussion
1. It sounds like you are looking for some area to direct your energy, and accounting is on the top of your list.	2.8 Attends to the surface content and feelings of the client. Supports client's questioning as having value, and hence client has value. Open to further dialogue.
2. Where have you been for the last three years? You are a poor lost soul.	1.0 Counselor ridicules the client, shows lack of caring, and ignores the client's feelings.
3. If you don't know what the opportunities are in the field of accounting, you might like to	2.0 Gives advice without understanding the situation. Attends only to the content of the client's

see someone in the business department or check the *Occupational Outlook Handbook.*	statement and doesn't deal with feelings. Does nothing to involve self in client's concern.
4. Graduation is a few months away, and you're scared that you will get out of here without any place to go. It sounds like accounting is an occupation that you may be vaguely interested in.	3.5 Goes beyond statement by attempting to attend to underlying message. Solicits action response on part of client.

CLIENT 2, A 25-YEAR-OLD MAN: I'm exhausted. My work has been building up for weeks. The harder I try to finish, the more there seems to be to do. I'm cranky with my wife and baby. I don't know what to do.

Responses	*Rating and Discussion*
1. Your job is getting to be too much, and you're worried about the effect it will have on your family.	2.5 Paraphrases statements and hence shows attentiveness to client's concern.
2. This will pass. Once part-time help is employed, things will ease off. Don't worry.	1.0 Responds by giving cheap advice. Behaves in manner congruent with preconceived role.
3. You feel you've hit the bottom of the pit. You work hard, but there is no end; you love your family but fight with them. And you sure don't want to stay there.	4.0 Goes beyond client statement. Designed to elicit reaction to the underlying message.
4. What do you want me to do? I have enough problems of my own.	1.0 Shows total lack of caring. Ridicules client by ignoring his appeal for help.

CLIENT 3, A COLLEGE FRESHMAN: You know, lately everything seems to be going wrong. My grades are slipping, my mother keeps hassling me, and my boyfriend has been threatening to break up with me.

Responses	*Rating and Discussion*
1. You know, I had the same problem when I was your age. . . . Don't worry, it will go away.	1.0 Gives advice without really understanding the problem, and an inappropriate self-disclosure.

2. You're having difficulty dealing with your mother and your boyfriend and keeping up your grades.

2.5 Attends to content by reflecting the main points made by the client.

3. You're very upset. You feel like you're being pulled in three different directions by your school, your mother, and your boyfriend.

3.5 Attends to feelings and content. Encourages client to go further by naming the three directional pulls.

4. It's very hard for you to decide what's best for you. And the pain only complicates the issue.

4.0 Goes beyond client statement; encourages client to react to deeper level feelings.

Practice Attending Exercises: "What Do I Do After I Say 'Hello'?"

The following exercises have been designed to enhance your attending skills. The discrimination exercises are designed to give you the opportunity to identify the client's underlying message, recognize various types of attending responses, and gain experience in learning how to distinguish effective from ineffective responses. The communication exercises offer you the chance to practice giving a variety of high level attending responses. Finally, the solo exercises provide further opportunities to develop your attending skills.

DISCRIMINATION EXERCISES. For each of the following excerpts identify the feeling and the content underlying the client's statement, and then indicate the *type* of attending response illustrated. Finally, point out the appropriateness of the response by rating the *level* of the response. Please be prepared to share the reasons for your ratings with your classmates and your supervisor. You may find that one or more of your classmates read the excerpts with another intonation than you did and consequently rated the responses quite differently.

CLIENT 1, A HIGH SCHOOL SOPHOMORE: I just got my report card, and I flunked geometry. I don't know how I'm going to face my parents. They're planning on a trip to Europe this summer, and now I'll have to go to summer school.

Underlying message: _____

Responses

Type and Level of Response

1. It's painful to disappoint your parents, but even more painful when you disappoint yourself.

1. _____

2. How am I going to face my   2. _____
   parents?

3. You seem to be upset for three   3. _____
   reasons. First, because you failed
   school; second, because you
   upset your parents; and third,
   because you might miss a trip to
   Europe.

4. Wow!   4. _____

CLIENT 2, A 30-YEAR-OLD MAN: I'm thinking about getting a divorce. My wife and I do nothing but tear each other to pieces. Whatever love we had for each other has long since been destroyed.

Underlying message: _____

Responses   *Type and Level of Response*

1. What was once a beautiful   1. _____
   relationship has deteriorated and
   become so painful that you feel
   there is only one way out.

2. Uh-huh.   2. _____

3. Being torn or tearing—it's so   3. _____
   painful.

4. You seem to be living painfully   4. _____
   with your wife. Your love for
   each other has gone, and you
   are wondering if divorce is the
   only sensible thing to do.

CLIENT 3, A 40-YEAR-OLD WOMAN: Ever since my breast surgery I have no desire to have sex with my husband. He's gentle and patient and doesn't mind looking at my scar. I have sex with him, but there's no feeling.

Underlying message: _____

Responses   *Type and Level of Response*

1. It's difficult to adjust to a pain- 1. _____
   ful experience. You feel that your
   relationship with your husband
   will never be the same again.

2. Even though you appreciate   2. _____
   your husband's efforts to be
   sensitive, you feel as though
   you've lost your desire for sex.

3. You have no feeling . . .   3. _____

4. Your adjustment to your breast   4. _____
   surgery has dimensions you
   never anticipated.

CLIENT 4, A HIGH SCHOOL JUNIOR: My older brother thinks he's my father. Do you know what he did? He went to the bar where I go once in a while for drinks with my friends, and he told the bartender not to serve me anymore because I'm not twenty-one.

Underlying message: _____

Responses   *Type and Level of Response*

1. You are mad at your brother   1. _____
   because he embarrassed you,
   and because you can no
   longer be served in that bar.

2. You resent someone else   2. _____
   running your life. You wonder
   who is going to be in charge—
   you or someone else.

3. Your big brother playing   3. _____
   "father's role" is really upset-
   ting to you.

4. He told him not to serve   4. _____
   you . . .

CLIENT 5, A 30-YEAR-OLD HOMEMAKER: I don't know what I want to do with my life.

Underlying message: _____

Responses   *Type and Level of Response*

1. You sound frustrated because   1. _____
   you want to change, but you
   are unsure of where to go
   or what to do.

2. You're confused, you feel you   2. _____
   want to do *something* with your
   life.

3. Thinking about the future is   3. _____
   scary.

4. I understand.   4. _____

CLIENT 6, A MIDDLE-AGED MAN: You know, I work hard all day. . . . I have lots of pressure. . . . People make me feel uptight. . . . Can I help it if I blow off a little steam in my wife's direction once in a while.

Underlying message: _____

Responses	Type and Level of Response
1. Something at work gets you so angry that when you get home you take it out on your wife.	1. _____
2. In your wife's direction . . .	2. _____
3. Continue.	3. _____
4. You really are burdened at this moment. Something seems to be making you uptight.	4. _____

CLIENT 7, A 19-YEAR-OLD MAN: The whole white society is corrupt. They're just out to manipulate and exploit the brothers, and that's just the way it is.

Underlying message: _____

Responses	Type and Level of Response
1. It's damning when everything is against you.	1. _____
2. You're angry at the way you feel white society treats you.	2. _____
3. The whole white society is corrupt.	3. _____
4. Damn.	4. _____

CLIENT 8, A YOUNG MAN: I don't know why the courts suggested I come and see you. So I got busted—that's no big deal. People get busted every day. The courts are always sending us to see this one or that one; they think I'm sick in the head. But I'm not, I'm not. I'll show them! I'll show them!

Underlying message: _____

Responses	Type and Level of Response
1. It's tough to be on the bottom. You'll climb the ladder to the top and tell them off.	1. _____
2. You're angry. The courts sent you here. You think it's because	2. _____

you're not functioning the way
they want you to.

3. I see, they sent you to me.	3. _____
4. And I'm not sick in the head.	4. _____

CLIENT 9, A MIDDLE-AGED WOMAN: This past weekend my cat died. She drank a can of paint thinner I left in the kitchen after I finished painting. I don't know how she got into the can. . . . She must have knocked the top off. I feel terrible now. . . . Oh, I just can't seem to do anything right.

Underlying message: _____

Responses

Type and Level of Response

1. It's awful to lose your pet,
    particularly one that you have
    been so attached to. But it is
    even more painful to have this
    feeling that you can't do
    anything right.

1. _____

2. It's hard when a pet dies,
    particularly if you feel you're
    partly at fault.

2. _____

3. I can't seem to do anything
    right. . . .

3. _____

4. Your cat's dying is painful, but
    it's more bothersome to you
    because you feel you can't do
    anything right.

4. _____

CLIENT 10, A NEWLY MARRIED WOMAN: When I get home from work, I just can't stand doing the housework anymore, especially washing the dishes. I'm so sick of them—I'm just letting them pile up in the sink! And I feel so guilty when my husband yells at me to do them. That's all he seems to be doing now—yelling!

Underlying message: _____

Responses

Type and Level of Response

1. It appears that you have
    ambivalent feelings about your
    relationship. Your husband con-
    stantly yells at you, and yet you
    seem to provoke him by not do-
    ing the dishes.

1. _____

2. You have to do the housework
   when you come home, a chore
   you hate, and your husband
   fails to understand you.

   2. _____

3. You are upset at your husband's
   yelling at you. He doesn't
   understand the pressure you
   have on the job.

   3. _____

4. You can't stand doing housework when you come home, so
   you don't do it. Your husband
   yells at you and makes you feel
   guilty.

   4. _____

COMMUNICATION EXERCISES. For each excerpt given below, give three different types of high level attending responses.

> *Client 11, a 30-year-old man:* I really love my work, but the foreman is really getting to me. He used to be my best friend, but he's changed. I thought he knew a lot, but now I see that he's not very good at his job.
>
> *Client 12, a middle-aged woman:* I have a difficult problem. I went back to work after being home for many years. Now I'm working and making much more money than my husband, and it's tearing him apart. I don't know what to do.
>
> *Client 13, a 40-year-old woman:* I'm beginning to feel very useless . . . no one needs me anymore. My husband and I don't seem to communicate. I don't see my children that much—they have their own friends and interests. Most of the time I'm alone.
>
> *Client 14, a 25-year-old man:* My girlfriend has cancer, and it's driving me crazy. The doctors say it's terminal. I don't have much faith in doctors. If anything happens to her, I don't know what I'll do . . . I love her so much.
>
> *Client 15, a 40-year-old man:* I'm overwhelmed by the amount of work I have. My wife died five years ago, and my son is on drugs. I don't know what to do.
>
> *Client 16, a 24-year-old woman:* I'm really having a bad time at home. I just wish I weren't there. I'm twenty-four years old, and my parents expect me to act like a child. They just don't understand much about the world. I don't know what to do about it.
>
> *Client 17, a high school junior:* My grades are so bad I'm sure I won't be able to make any of the colleges I want to go to.
>
> *Client 18, a 30-year-old woman:* I feel very depressed lately. I have a good job, but it's not enough. My family thinks I should get married, but I don't know anyone I'd like to marry.

Client 19, a 25-year-old woman: I'm so hurt. We've been married only six months, and John wants a divorce. He has completely rejected me. I don't know where I failed.

Client 20, a college junior: I don't know whether I should go to law school or not. I'd like to be a lawyer, but everyone tells me how tough law school can be. I'm afraid of flunking out.

Solo exercises. Practice your attending responses in a triad. Each counselor-in-training should take a turn at being a "client," a "counselor," and a "supervisor." Role-play on a tape recorder. After 5 minutes the "supervisor" should lead the critique of the counselor's use of the attending role. Repeat the exercise until all members of the triad have had the opportunity to practice.

The Clarifying Role

This role is employed when the counselor actively attempts to make various issues clearer or to dispel any confusion that may exist. As the counselor, you may be unsure of what the client is actually trying to say or what the client means by his or her statement; or you may feel that the client is confused and needs help in clarifying an issue; or you may believe that the client needs to provide more information or elaborate on a problem or concern (Cormier & Cormier, 1991). Egan (1990) points out that it is important to understand what clients are attempting to express rather than feign understanding. The clarifying role is often used in conjunction with the attending role because it can facilitate the understanding of the client's thoughts and feelings.

Your clarifying response should encourage your client to reflect, restate, or redefine a statement he or she has made about a situation, problem, or concern. The request for clarification may be communicated nonverbally by silence or by a quizzical look; paraverbally by an inquisitive tonal quality; or verbally by an open-ended question that cannot be answered by a simple yes or no, or by a statement that reflects your perception of the issue.

Types of Clarifying Responses

The clarifying role generally includes the following types of responses:

1. *Perception checking.* In this response the counselor seeks to verify the accuracy of his or her perception of all or part of the client's message, and the counselor asks the client to confirm or modify this perception (Brammer, 1988). A typical perception check is "It seems your major concern is . . ."

2. *Clarification of alternatives.* This response is used when the counselor has some doubt or confusion about which alternative is of major importance to the client or when the counselor believes the client is not sure of what he or

she really wants to focus on. Often this response takes the form, "Do you mean this or do you mean that?"

3. *Request for further information or elaboration.* In this response the counselor makes a specific request for further information in order to clear up any confusion that may be present. A response of this sort is, "I'm not sure what you mean, please go on."

Illustrations of Different Types of Clarifying Responses

The following examples illustrate the use of the clarifying role.

CLIENT 1, A 20-YEAR-OLD WOMAN: I feel as if my friends have been letting me down lately. Until recently I was very popular in school—I had a lot of friends, and I enjoyed being with them. But all of a sudden I feel like I've been losing them. I've been snubbed by some, and I really don't know why this is happening.

1. *Perception checking:* I gather your major concern is looking at yourself to see what you've done to cause your friends to pull away.
2. *Clarifying alternatives:* I'm not sure what your real concern is. Are you angry because you may have done something to lose some of your friends, or do you feel you've done nothing and can't understand it?
3. *Request for further information or elaboration:* I'm not sure I completely understand your concern. Tell me more about your situation.

CLIENT 2, A TEENAGER: My mother is constantly on my back. Every time I turn around, there's something else she wants me to do or some more advice she wants to hand out. She really burns me up.

1. *Perception checking:* Are you saying your mother is controlling your whole life?
2. *Clarifying alternatives:* Is it really your mother who is bothering you, or is it your own lack of skill in speaking up for your rights that bothers you?
3. *Request for further information or elaboration:* I'm uncertain about your relationship with your mother and other family members. Please amplify this.

CLIENT 3, A 23-YEAR-OLD WOMAN: I'm working part time as a dental assistant; it's the only job I could get. It seems that I've spent my whole life wanting and training to be a teacher, but it's impossible to get a job. I hate what I'm doing now, and I'm so afraid I'll be stuck there for the rest of my life and my education will just go down the drain.

1. *Perception checking:* You seem to feel that the most important thing for you is to get another job, either in education or in a closely related field.
2. *Clarifying alternatives:* You seem to be very angry, but I'm not sure whether that anger is directed toward yourself for not adjusting to this situation or toward others who may have encouraged you.
3. *Request for further information or elaboration:* I hear an angry voice, and I'm puzzled about whom this anger is directed toward. Please continue.

Levels of Clarifying Responses

This role is often used in the early phases of the counseling process. High level responses are communicated by words and by paraverbal and non-verbal means that reveal a sincere and genuine interest in the client and focus on critical issues or situations that need clarification. The effective use of this role greatly enhances both the client's and your understanding of issues stated by the client or within the client's internal frame of reference. It can be ineffective and even counterproductive when you employ incongruous paraverbal or verbal channels, constantly attempt to clarify some trivial points, show approval or disapproval, or use the role excessively. The excessive use of this role is distracting to clients and tends to interrupt their thought processes (Hackney & Cormier, 1988). The use of this role can be rated on the four-point scale indicated below.

Level 1: Poor use of the role. The counselor is most ineffective when he or she reveals little or no interest in the client's underlying concern, communicates a value judgment in the response, or uses the role at an inappropriate point in the counseling process. Low level responses reveal little relationship to the feeling or the content expressed by the client, and the client may reveal a preoccupation with other matters.

Level 2: Mediocre use of the role. The counselor is not very effective when he or she tries to clarify only a part of the client's underlying message; when the response is a superficial interpretation of the client's verbal, paraverbal, and nonverbal message; and when only minimal interest in the client is revealed.

Level 3: Good use of the role. The counselor is effective when he or she focuses on the stated content and feelings of the client; attempts to clear up garbled, confused, or unclear messages; and manifests a high degree of interest in the client.

Level 4: Excellent use of the role. The counselor is most effective when he or she goes beyond the stated content and surface feelings expressed by the client; helps the client articulate issues more clearly; assists the client in focusing on the underlying concerns; avoids value judgments; and shows an intense interest in the client.

Illustrations of Different Levels of Clarifying Responses

To give you some practice in learning to distinguish between effective and ineffective counselor responses, three client statements are presented below, and each statement is followed by four different clarifying responses. The Level-of-Response Scale was used to rate each of the counselor's responses, and the reasoning behind each of these ratings is stated. In studying these examples, you may find that you do not agree with these ratings. Because tonal quality and other paraverbal aspects of the client's statements and the counselor's responses are not conveyed, it is possible that you will read some of these statements in another way than the person who rated them. Any response may obtain a different rating depending on the paraverbal message that the reader hears. Thus, it is crucial to indicate the reasons behind the ratings so that any differences among these ratings can be understood.

CLIENT 1, A HIGH SCHOOL SENIOR: I'm worried about what I'll do in the future, and I was wondering whether you can tell me if there are any good jobs in accounting.

Responses	*Ratings and Discussion*	
1. I'm not sure what your real concern is. Are your unsure about our future in general or unsure about accounting as an occupation?	3.0	The counselor goes beyond the stated message and attempts to clarify between two reasonable alternative meanings.
2. Your major concern seems to be whether or not you should go on to college and major in accounting.	2.8	The counselor is checking whether or not his perception of the client's underlying message is correct.
3. Is the concern about your future coming only from you, or is there pressure from home?	2.8	The counselor clarifies between alternatives to see if there is another issue in the client's presenting problem.
4. Are your grades good enough to get into a good accounting program?	1.5	The counselor attempts to have the client elaborate, but communicates a value judgment in the response.

CLIENT 2, A 25-YEAR-OLD MAN: I'm exhausted. My work has been building up for weeks. The harder I try to finish, the more there seems to be to do. I'm cranky with my wife and baby. I don't know what to do.

Responses	*Ratings and Discussion*	
1. You seem to be saying that the pressure of the job is causing problems with your health and family life.	2.5	The counselor checks on her perception of the client's statement. Does not go much beyond surface level.

2. I'm unclear about your relationship with your boss and your work; please tell me about that situation.

2.8 The counselor requests the client to elaborate. Focus is on one area that is reported by the client.

3. I'm not sure what you mean. Is your health affecting your work, or is it vice versa?

3.0 The counselor is clarifying between two important alternatives.

4. You seem frustrated and think that the cause lies in your work, but you are uncertain if that is the true cause.

3.5 The counselor goes beyond the stated message in order to clarify the underlying causes of the client's feelings.

CLIENT 3, A COLLEGE FRESHMAN: You know, lately everything seems to be going wrong. My grades are slipping, my mother keeps hassling me, and my boyfriend has been threatening to break up with me.

Responses

1. You feel you'd like to stop the world and start all over again.

Ratings and Discussion

2.5 The counselor checks his perception on the seriousness of the problem. May sound too flippant for some clients.

2. You seem to feel that everyone is against you.

3.0 The counselor's perception check goes beyond the problems stated by the client.

3. I'm not sure which one irks you more—your poor school work, your relationship with your mother, or your boyfriend.

2.5 The counselor's clarification among the stated alternatives may sound a bit superficial to the client.

4. And you're wondering whether you should quit school, move out of the house, or get another boyfriend.

3.0 By injecting humor into the clarification among alternatives, the counselor is attempting to go beyond the surface meaning of the client's statement.

Practice Exercises: "What Do I Say After I Say 'I See'?"

These exercises have been prepared to assist you in developing your clarifying skills. The discrimination exercises will give you the chance to further practice identifying the client's underlying messages, recognizing the different ways that client statements may be clarified, and discriminating between more and less effective responses. The communication exercises will enable you to practice giving high level clarifying responses in a variety of client situations. Finally, the solo exercises will give you the chance to sharpen both your attending and clarifying role communication skills.

DISCRIMINATION EXERCISES. For each of the following excerpts identify the feeling and the content of the client's message, and then specify the type of clarifying response used; finally, rate the level of the response. Be ready to discuss the reasons for your ratings with your classmates and your supervisor. You may find that the excerpts were read differently by one or more of your classmates; thus, they may have given another rating to the response.

CLIENT 1, A HIGH SCHOOL SOPHOMORE: I just got my report card, and I flunked geometry. I don't know how I'm going to face my parents. They're planning on a trip to Europe this summer, and now I'll have to go to summer school.

Underlying message: _____

Responses	*Type and Level of Response*
1. I'm not sure why you're upset. Is it because of failing, missing the trip to Europe, or disappointing your parents?	1. _____
2. Are you fearful that your parents might punish you because you upset their plans?	2. _____
3. Please tell me more. I'm a bit vague about what you're most fearful about.	3. _____

CLIENT 2, A 30-YEAR-OLD MAN: I'm thinking about getting a divorce. My wife and I do nothing but tear each other to pieces. Whatever love we had for each other has long since been destroyed.

Underlying message: _____

Responses	*Type and Level of Response*
1. I'm not sure what you're saying. Is it, "It's too late to patch this marriage up" or, "I'd really like to try to bring the relationship back to where it was"?	1. _____
2. What hurts the most—the fighting, the pain, or the loss of the relationship?	2. _____
3. I sense some ambiguity in what you're feeling. Please tell me more.	3. _____

CLIENT 3, A 40-YEAR-OLD WOMAN: Ever since my breast surgery I have no desire to have sex with my husband. He's gentle and patient and doesn't mind looking at my scar. I have sex with him, but there's no feeling.

Underlying message: _____

Responses	*Type and Level of Response*
1. Are you upset because you're having difficulty adjusting to the surgery, or because you can't talk to your husband about it?	1. _____
2. You seem to be saying that your reaction to your breast surgery has affected your relationship with your husband more than anything else.	2. _____
3. I'm somewhat uncertain about your adjustment to your surgery. Go on.	3. _____

CLIENT 4, A HIGH SCHOOL JUNIOR: My older brother thinks he's my father. Do you know what he did? He went to the bar where I go once in a while for drinks with my friends, and he told the bartender not to serve me anymore because I'm not twenty-one.

Underlying message: _____

Responses	*Type and Level of Response*
1. I'm not sure why you're mad at your brother. Is it because of this one incident, or does it go beyond this?	1. _____
2. You feel that you do not have a good relationship with your brother.	2. _____
3. I'm a bit confused about your relationship with your brother. Please tell me more about that.	3. _____

CLIENT 5, A 30-YEAR-OLD HOMEMAKER: I don't know what I want to do with my life.

Underlying message: _____

Responses	*Type and Level of Response*
1. It sounds like you want a change in your daily routine.	1. _____

It seems you might want to go
back to work.

2. I'm not sure what you mean. Is   2. _____
it that you don't have a road
map, or is it that you have the
map but don't know what direction to go in?

3. You sound a little overwhelmed,   3. _____
and I'm vague about why this
is. Tell me more about yourself
and your situation.

CLIENT 6, A MIDDLE-AGED MAN: You know, I work hard all day. . . . I have
lots of pressure. . . . People make me feel uptight. . . . Can I help it if I blow
off a little steam in my wife's direction once in a while?

Underlying message: _____

Responses   *Type and Level of Response*

1. You seem to be saying that   1. _____
things are a mess at home, and
it's because of the pressure from
your job.

2. I'm uncertain about your real   2. _____
source of anger. Is it the boss,
the system, or the whole ball of
wax?

3. I'm not clear on the point   3. _____
you're making. Please help me
by describing your life situation
in more detail.

CLIENT 7, A 19-YEAR-OLD MAN: The whole white society is corrupt.
They're just out to manipulate and exploit the brothers, and that's just the way
it is.

Underlying message: _____

Responses   *Type and Level of Response*

1. Is it really the whole white so-   1. _____
ciety that you're upset about, or
did a recent incident upset you?

2. You seem to be blaming society   2. _____
for something. I'm not sure
what that something is.

3. Go on.   3. _____

CLIENT 8, A YOUNG MAN: I don't know why the courts suggested I come and see you. So I got busted—that's no big deal. People get busted every day. The courts are always sending us to see this one or that one; they think I'm sick in the head. But I'm not, I'm not. I'll show them! I'll show them!

Underlying message: _____

Responses	*Type and Level of Response*
1. Are you angry because you had to come to see me, or are you upset because others are in control of your life?	1. _____
2. You are really very upset. It sounds like you're angry at yourself because you got caught.	2. _____
3. You sound very angry. But I'm not sure who you are really mad at. Please continue.	3. _____

CLIENT 9, A MIDDLE-AGED WOMAN: This past weekend my cat died. She drank a can of paint thinner I left in the kitchen after I finished painting. I don't know how she got into the can. . . . She must have knocked the top off. I feel terrible now. . . . Oh, I just can't seem to do anything right.

Underlying message: _____

Responses	*Type and Level of Response*
1. You're so angry at yourself because of your cat dying that you feel you can't do anything but cry now.	1. _____
2. Is it the loss of the cat that has hurt you, or is it really something else?	2. _____
3. I'm a bit puzzled. The cat's death seems to have set you off. Please tell me more about yourself.	3. _____

CLIENT 10, A NEWLY MARRIED WOMAN: When I get home from work, I just can't stand doing the housework anymore, especially washing the dishes. I'm so sick of them—I'm just letting them pile up in the sink! And I feel so guilty when my husband yells at me to do them. That's all he seems to be doing now—yelling!

Underlying message: _____

Responses	*Type and Level of Response*
1. I'm not sure who you're really angry at—your husband or yourself?	1. _____
2. Is it the pressure from work that's getting to you, or is there something at home that's causing this problem?	2. _____
3. Several things appear to be bothering you, perhaps some more than others. Let's talk about them.	3. _____

COMMUNICATION EXERCISES. For each excerpt given below please give three different types of high level clarifying responses.

Client 11, a 30-year-old man: I really love my work, but the foreman is really getting to me. He used to be my best friend, but he's changed. I thought he knew a lot, but now I see that he's not very good at his job.

Client 12, a middle-aged woman: I have a difficult problem. I went back to work after being home for many years. Now I'm working and making much more money than my husband, and it's tearing him apart. I don't know what to do.

Client 13, a 40-year-old woman: I'm beginning to feel very useless . . . no one needs me anymore. My husband and I don't seem to communicate. I don't see my children that much—they have their own friends and interests. Most of the time I'm alone.

Client 14, a 25-year-old man: My girlfriend has cancer, and it's driving me crazy. The doctors say it's terminal. I don't have much faith in doctors. If anything happens to her, I don't know what I'll do . . . I love her so much.

Client 15, a 40-year-old man: I'm overwhelmed by the amount of work I have. My wife died five years ago, and my son is on drugs. I don't know what to do.

Client 16, a 24-year-old woman: I'm really having a bad time at home. I just wish I weren't there. I'm twenty-four years old, and my parents expect me to act like a child. They just don't understand much about the world. I don't know what to do about it.

Client 17, a high school junior: My grades are so bad that I'm sure I won't be able to make any of the colleges I want to go to.

Client 18, a 30-year-old woman: I feel very depressed lately. I have a good job, but it's not enough. My family thinks I should get married, but I don't know anyone I'd like to marry.

Client 19, a 25-year-old woman: I'm so hurt. We've been married only six months, and John wants a divorce. He has completely rejected me. I don't know where I failed.

Client 20, a college junior: I don't know whether I should go to law school or not. I'd like to be a lawyer, but everyone tells me how tough law school can be. I'm afraid of flunking out.

SOLO EXERCISES. Practice your attending and clarifying responses in a triad. Be careful not to use any other type of response. Give all members in the triad a chance to practice their responding skills. Tape record your sessions so you can review why some responses were better than others.

THE SUPPORTING OR REASSURING ROLE

Communicating support is an important relationship skill (Johnson, 1990). Everything that a counselor does to build the foundation for a solid counseling relationship is a form of support and reassurance. Your attitude and effective use of attending and clarifying responses demonstrate support by showing a real interest in the client. The specific skills discussed in this role are used when you want to convey this support and reassurance in a more concrete way. This role involves actively providing positive feedback; communicating feelings of security, reassurance, and encouragement; and reaffirming the client's sense of self. In effect, you are saying "I like you" or "You are okay" to the client. This role may be explicitly or implicitly communicated by nonverbal and paraverbal, as well as verbal, means. This role is often necessary in a counseling relationship because clients usually need to feel accepted, and they are frequently unable to marshal their own strengths to engage in new behaviors, to bring about desired changes, or to find appropriate solutions to their problems without the counselor's support.

The supporting or reassuring role is used when you want to acknowledge the experience of the client as very real and to make the client feel that he or she is heard, understood, and accepted as a person of value even though his or her behavior or specific actions may not be acceptable or liked by you or significant others. Benjamin (1987) suggests that reassurance responses help clients overcome blockages and deal with difficult problems and situations, and Brammer (1988) points out that supportive responses are used with clients who are grief stricken or in a state of crisis.

Supportive and reassuring responses should stress that you have faith in the client; believe in the client's ability to resolve issues; have an understanding of the frailty of the human condition; and respect the dignity and worth of the client. In using supportive responses, you should avoid comparisons with others and focus on the idea of being and doing rather than on the attainment of a prize or another status symbol. The role is employed at various stages in the counseling process.

Types of Supporting or Reassuring Responses

Typical responses that demonstrate this role are:

1. *A person-of-value response.* This type of response is designed to show unconditional respect. It communicates to clients that they are individuals of significant worth and value regardless of their actions. Responses of this sort focus on the uniqueness of the person; the intrinsic aspects of life rather than the extrinsic aspects; the person's existence rather than the person's productivity; the effort rather than the end product; and *what* one is doing rather than *how* one is doing. "You seem like a fine person to me" emphasizes one's uniqueness. And phrases such as "You handled that nicely" and "It sounds like you had fun" separate worth from accomplishment and can provide encouragement and solid humanistic support to clients.

2. *An approval response.* This sort of response is used to help clients feel more positive about themselves. This is accomplished by showing approval, agreement, or reassurance for a particular thought, feeling, or behavior that the client has revealed. The response may applaud some previous event, compliment some present activity, or reassure that a future event will turn out appropriately. Phrases such as "It was nice for you to help Mary," "You look nice today," and "You have the ability to do that" praise the person and tend to make the client feel better.

3. *A consolation response.* Consolation responses communicate a caring concern and show clients that you are supportive of them when they are not feeling good about something. This kind of response can be very useful when a client has gone through a recent unhappy occurrence, is currently dealing with an uncomfortable situation, or is facing an unpleasant future event. The response may be an expression of comfort, sympathy, or condolence. Phrases such as "What an awful experience for you," "You have my sympathy," and "You are really facing a tough situation" can show your understanding of the client's problem and your support for the client.

4. *A relaxation response.* Relaxation responses can be very supportive when clients are tense, excited, and overstimulated. They are designed to get clients to pause, to become calmer, and to become more in touch with all their emotions. Phrases such as "Take a deep breath," "Close your eyes for a few minutes," and "I want you to think about a pleasant scene" can communicate a sincere interest in the client.

Illustrations of Different Types of Supporting or Reassuring Responses

The following three examples illustrate the use of the supporting role.

CLIENT 1, A YOUNG WOMAN: My father and I fight all the time, especially about the boys I know. He won't even let a boy in the house. He says they're all no good. I think he just doesn't trust me.

1. *A person-of-value response:* You really are a spunky young lady. You are willing to stick up for the things you believe in.
2. *An approval response:* It is really hard for you to feel that you're not trusted. It's very good to tell your father how you feel about your rights.
3. *A consolation response:* I can see you're very angry. It's really hard when your father doesn't understand your position and comes down hard on you.
4. *A relaxation response:* Wow, that's a tough spot to be in. You're very excited. Before we begin, I want you to sit down and think about something pleasant for 30 seconds.

CLIENT 2, A MIDDLE-AGED MAN: All my life I've never been able to hold a job for very long. Everything goes okay for a while, but then I get into a fight with the boss or something like that and I'm fired. I can't seem to help it.

1. *A person-of-value response:* Even though you may have had some problems with some jobs you've held, you seem to have some wonderful qualities.
2. *An approval response:* Yet, every time that you spoke up I'm sure you felt there was a good reason for it.
3. *A consolation response:* It's really tough not being able to hold on to a job for a long time.
4. *A relaxation response:* I want you to tell me some more about yourself, but first, I want you to take a deep breath and systematically relax different parts of your body.

CLIENT 3, A MIDDLE-AGED WOMAN: My mother died a few months ago, and I haven't been the same since. I just can't seem to get my life back on track again. I've been very preoccupied with death and dying.

1. *A person-of-value response:* This emotional reaction tells me that you were a wonderful daughter. After the death of your mother, it's normal to feel derailed.
2. *An approval response:* Even though being unable to think or act clearly is upsetting, your mourning your mother's death is a healthy response. Your feelings are appropriate. Allow yourself time to grieve.
3. *A consolation response:* I'm sorry to hear about her death. The death of a parent is painful and a big loss for you.
4. *A relaxation response:* Mourning takes time, and unfortunately, you are not going to be able to turn it off or rush it. When the thoughts of death and dying come, remember that they are letting you know how much you loved your mother and to value the people in your life more fully. You need to take time to reflect on this.

Levels of Supporting and Reassuring Responses

Counselors who use this role skillfully show respect for their clients, reveal an understanding of the strengths and weaknesses of human beings, normally employ a relaxed warm tone in their voice, and use attentive body gestures. The effective use of the role can help clients reduce the intensity of their feelings, lower their anxiety about some activity, feel more secure in the counseling relationship, and gain a sense of confidence. These responses should help clients mobilize their strengths by allowing them to put the past into perspective, explore alternatives, and face the future positively. Support and reassurance help reinforce desired behavior. The role is ineffective and may be counterproductive when the counselor fails to acknowledge the depth of the client's feelings or the seriousness of the client's concerns, and when it is used in inappropriate ways. This can occur when the counselor sounds insincere, downplays the seriousness of the client's problem, overuses the role, or fosters a dependency relationship, or when there is real danger that the counselor's words could be interpreted as cheap advice rather than support. Such phrases as "Don't worry, things will turn out okay" convey disinterest or lack of understanding rather than support or reassurance. The timeliness and appropriateness of the counselor's explicit use of support is a function of the immediacy of the client's need for reassurance.

Similar to other counselor role responses, the supporting and reassuring role may be used with various degrees of effectiveness. Here is a four-point scale to rate the counselor's use of this role.

> *Level 1: Poor use of the role.* The counselor is ineffective in using this role when reassurance is inopportune, when it is conveyed in a sterile or condescending voice, when gestures are distracting, or when the counselor is insincere or fails to show meaningful support. Ritualistic reassurance, including clichés, may be used.
>
> *Level 2: Mediocre use of the role.* The counselor is somewhat ineffective when support is given in a lukewarm manner, when it is poorly phrased, or when the role is of limited value at the point it is used in the counseling process.
>
> *Level 3: Good use of the role.* The counselor is effective when he or she is responding appropriately to the needs of the client, reassuring or supporting properly, and using this role at a suitable time in the counseling process.
>
> *Level 4: Excellent use of the role.* The counselor is most effective when the counselor's tone and words are very supportive, and when the counselor's use of the role is an appropriate response to the client's needs.

Illustrations of Different Levels of Supporting or Reassuring Responses

The following illustrations are provided so that you can learn to perceive the differences between effective and ineffective use of the supportive and reassurance role. In each of the examples, the responses have been rated using the Level-of-Response Scale and the reasons for assigning these ratings is reported. Examine each of these examples to see if you agree with the ratings and the reasons stated. If you do not agree with these ratings, you may have read one or more of these statements with another intonation than the person who rated them. Because the paraverbal aspects of the client's statements and the counselor's responses are never adequately indicated by the printed word, any response may be rated quite differently depending on how it is heard. It is very important that you give reasons for your ratings so that you can discuss any differences with your supervisor and classmates.

CLIENT 1, A COLLEGE SENIOR: I'm so mad at that dean. He just asked why I took part in that demonstration. He acted as if I did something wrong.

Reponses	*Ratings and Reasons*
1. You are a person who acts according to your principles. That's a fine quality.	2.0 This person-of-value response can be misleading. The client may perceive it as support for his behavior.
2. Wow! You sound so agitated— he really got to you. I want you to sit down, take a deep breath, and relax.	3.0 This relaxation response should enable the client to become more relaxed and calmer.
3. You felt justified in participating in the demonstration, and it hurts when someone in authority questions actions that you think are most important.	3.5 The counselor shows approval for the client's feelings and support for acting in concert with one's principles.
4. Next time there is a demonstration, let me know and I'll go with you.	1.0 Misguided approval and inappropriate support.

CLIENT 2, A COLLEGE SENIOR: John and I have had a serious relationship for over a year, and he suddenly tells me he's going to live in Chicago to take care of his father's business—he left just like that.

Responses	*Ratings and Reasons*
1. That's okay. If you need to cry, cry. It's okay.	3.0 The counselor shows approval for the client's deep feelings of hurt.
2. It's understandable how you can feel hurt and discarded after such an experience.	3.5 This consolation response reveals the counselor's understanding and concern.

3. I know you are hurting now, and you have a right to feel that way.

3.0 The counselor's approval response shows awareness of the client's emotions and the situation.

4. I know you can overcome this loss. You have great strengths and can make new friends.

1.5 Although the counselor attempts to demonstrate high regard for the client, it doesn't sound genuine and may hold out false hope.

CLIENT 3, A 35-YEAR-OLD WOMAN WHO RECENTLY LOST HER HUSBAND: It's only when I'm alone that these feelings and memories keep returning. If I don't do something right away to divert my attention, I just start crying and feeling sorry for myself.

Responses

1. Your feelings and memories are part of your life. They are telling you how much you love and miss your husband and how he loved you. There's nothing wrong with feeling sorry for yourself as you miss him.

Ratings and Reasons

3.5 The counselor's response shows appropriate approval for the client's feelings and behaviors.

2. You are a fine person. This battle going on between your head and your heart shows your respect for your husband and yourself.

3.5 This person-of-value response is combined with an approval response and shows respect for the client's deep feelings.

3. I really feel for the struggle you are experiencing while you're mourning your husband.

3.0 This response shows consolation for the client's feelings and loss.

4. I understand your desire to resume a normal life, but your grieving may be more important and take a while to resolve. It's important to allow yourself time to cry.

3.5 The counselor shows awareness of the client's needs and pressures, and the approval reveals her understanding of these dynamics.

Practice Supporting Exercises: "How Do I Say 'You're Okay'?"

The following set of exercises is designed to facilitate the development of your supporting and reassuring role skills. Similar to the previous sets of exercises, it contains three types of skill-building practices: discrimination exercises, communication exercises, and solo exercises.

DISCRIMINATION EXERCISES.   For each client statement given below, indicate the feeling and the content underlying the client's statement. Then identify the *type* of supporting or reassuring response employed, and show the appropriateness of the response by rating the *level* of the response. Anticipate your classmates' questions about the reasons for your answers, and be prepared to discuss your reasons in class. The excerpts may have been read differently by your classmates, and hence their ratings may not agree with yours. It is important to understand the reasons why these differences exist.

CLIENT 1, A MIDDLE-AGED WOMAN: My husband and I are very upset with our daughter. Every time she comes home it ends up in some sort of conflict. She's twenty-three years old and collecting unemployment. She doesn't seem to be going anywhere. I just wish she would get married.

Underlying message: _____

Responses	*Type and Level of Response*
1. You are a good mother to be concerned for her welfare.	1. _____
2. It's quite normal to be upset when children disappoint you.	2. _____
3. You're under a great deal of pressure. Take a deep breath and relax.	3. _____
4. It's hard to face these daily fights. I understand your desire to see her settle down.	4. _____

CLIENT 2, A YOUNG MAN: I think I want to get a divorce. When I married Susan I thought we had a lot in common. But I found out otherwise. Her values are so materialistic that I find them repulsive.

Underlying message: _____

Responses	*Type and Level of Response*
1. You're hurting now. You feel that the marriage is not working out the way you wanted it to.	1. _____
2. Wanting to live by values that you cherish is really wonderful. You are a fine person for wanting to do that.	2. _____
3. Right now you're angry with Susan and see only where you disagree. Before we start dis-	3. _____

cussing this issue, I want you to take a deep breath.

4. It really is difficult to live with   4. _____
someone who doesn't share
your values.

CLIENT 3, A MAN IN HIS LATE 50s: I just got laid off from my job. I worked for the place for eighteen years, and just like that I'm out on the street. It's impossible for a man my age to begin again.

Underlying message: _____

Responses   *Type and Level of Response*

1. After eighteen years, it's tough   1. _____
to be laid off. You were not
ready for a change.

2. You must have been a good em-   2. _____
ployee, having remained with
the same company for eighteen
years.

3. Sometimes we think that the   3. _____
cards in this life haven't been
dealt from a fair deck. The in-
justice gets to you.

4. How can a good man be kept   4. _____
down? It might be hard work to
find another job, but with your
tenacity I'm sure it can be done.

CLIENT 4, A YOUNG WOMAN, A HIGH SCHOOL SENIOR: I'm not sure where I'm at. I'd like to go to college and get married. I think raising a family can be a full-time commitment. But then again I should think of some career or occupation. Suppose I don't marry or can't have kids. I'm kind of muddling through.

Underlying message: _____

Responses   *Type and Level of Response*

1. The kinds of questions you're   1. _____
asking yourself are good ones,
even though there may not be
an immediate answer to them.

2. Women are spending more and   2. _____
more of their lives in the world
of work. So it's good that
you're thinking about college
and its vocational implications.

3. Hey, you're great. Many girls
   your age don't think about their
   future with such insight.

   3. _____

4. Women today can find fulfillment through a variety of
   options. I'm happy to see you
   examining them.

   4. _____

CLIENT 5, A 20-YEAR-OLD WOMAN: Everything was great when I first met
Tom. We had so much in common, and he seemed like the right guy to marry.
But lately we're fighting a lot. I'm starting to see a different side of Tom, and
I don't like it. I still really love him though. He's a great guy. Maybe we'll work
things out when we're married.

Underlying message: _____

Responses   *Type and Level of Response*

1. You're a very capable girl. I'm
   sure you'll be able to handle
   any problems that come up.

   1. _____

2. It's very common for people to
   see the other aspects of a person as they grow closer. It's
   good that you are seeing this
   side of Tom now.

   2. _____

3. The closer two people get, the
   more they have to learn to deal
   with their differences. It's okay
   to recognize that there are
   things you don't like in each
   other. It doesn't mean your relationship is good or bad . . . it
   just means you're both more
   aware.

   3. _____

4. It's normal to have doubts about
   your relationship. Before you
   say anything else, I want you to
   close your eyes and relax for 30
   seconds.

   4. _____

CLIENT 6, A 20-YEAR-OLD MAN: Last week I lost my job. My boss told
me to do something I didn't think was right. I wouldn't do it, and he fired me.
I don't know what I should do. I have a family to support.

Underlying message: _____

Responses	*Type and Level of Response*
1. I admire your willingness to maintain your principles. You were willing to lose your job. That took courage.	1. _____
2. You must be under some pressure. I know we can work things out for you and your family.	2. _____
3. You're a man with integrity and conviction. Many firms are looking for people with those characteristics.	3. _____
4. I know you are very concerned, but I don't see how you had much choice. After all, what the boss wanted you to do was illegal, unethical, or both.	4. _____

CLIENT 7, A YOUNG MAN, 16 YEARS OLD: There's this girl that I like, but I can't ask her out because my friends tell me that I'm too short and ugly, and no girl would go out with me. I don't know what to do. I really like this girl.

Underlying message: _____

Responses	*Type and Level of Response*
1. You are being very hard on yourself. You have many good qualities that many people admire.	1. _____
2. Some may say you're short and ugly, but I don't see you that way. To me you're a heck of a nice guy.	2. _____
3. It's natural to want to ask a girl out whom you admire. And unfortunately, it's also typical for friends to tease.	3. _____
4. I'm glad you came to see me. Let's think about various ways you can get to know this young lady better.	4. _____

CLIENT 8, A YOUNG WOMAN: If I could tell my boss, just once, where to get off, I know I'd feel better, but I know I'd lose my job. I don't think it would be worthwhile.

Underlying message: _____

Responses	*Type and Level of Response*
1. I know you'll find a creative way to deal with the situation. Let's talk about it for a while.	1. _____
2. When you find yourself holding in too much anger, find a way to relax and get the anger out—talk to a friend, jog, listen to relaxing music, or close your eyes and take a deep breath.	2. _____
3. You have a great deal of patience and determination in staying with your job. It's a sign of maturity as you seek a solution to your problem at work.	3. _____
4. Sometimes we have to choose to put up with frustration. It's an awful choice. I really feel for you.	4. _____

CLIENT 9, A COLLEGE SOPHOMORE: I just quit my job because I found it too hard to keep up my studies and work after school. By the time I got home I was too tired to study. My father thinks I'm copping out on my responsibilities, and he's really mad.

Underlying message: _____

Responses	*Type and Level of Response*
1. You surely felt that you did the right thing in quitting. You must have felt you had to quit the job or fail out of school.	1. _____
2. You are a fine person. You seem to have a considerable amount of good sense.	2. _____
3. Fathers can be tough.	3. _____

4. It's difficult to try to maintain   4. _____
   a good average and a job at the
   same time. You deserve some
   credit for trying to support
   yourself while you're in school.

CLIENT 10, A MIDDLE-AGED MAN: My son and I don't get along very well. I guess one could call it the generation gap, but I'm only trying to do what's best for him.

Underlying message: _____

Responses   *Type and Level of Response*

1. Being a parent is not easy.   1. _____

2. You are really concerned about   2. _____
   your son, and I'm sure you
   only have his best interests at
   heart.

3. Mark Twain once said that at   3. _____
   eighteen he thought his father
   was awfully stupid; but when
   he was twenty-one he couldn't
   get over how much his dad had
   learned in three years.

4. I know the generation gap is a   4. _____
   difficult one to bridge, but we
   can learn to span that gap fairly
   well. Before we begin I want
   you to take a deep breath.

COMMUNICATION EXERCISES. Please respond to each client statement in this set of exercises by furnishing at least two different types of high level supportive statements. Try to construct your responses as quickly as possible.

> *Client 11, a college freshman:* I came because I had a terrible feeling of depression. My boyfriend, who's nineteen, is threatening to break up with me because I won't have sex with him. We've been going together for sixteen months, and everything was great until about two months ago. Now he has this big thing about going to bed. He keeps saying I'd do it if I really loved him. He doesn't understand my viewpoint at all.
>
> *Client 12, the wife of an alcoholic husband:* My problem is that my husband is an alcoholic. I'm the sole supporter of my family, and although I loved him once, and suppose I still do, I don't think that staying with him is helping the children.

Client 13, a woman who has had three miscarriages: I've been so terribly depressed lately that I can no longer function effectively. My husband is considerate and understanding, but he says I must resign myself to the fact that I will never have children. I just can't accept that.

Client 14, a 10-year-old girl: Nobody likes me. I don't have any friends. My mother says not to feel bad. I can't help it if I want other kids to like me.

Client 15, a teenager: When I get home from school my mother is always drunk. The house is a mess. She doesn't do the dishes or clean up, and she hardly cooks any supper. She just lies there and drinks and screams at me, and I scream back. I'm so ashamed I can't ask my friends to come over, and I'm so disgusted sometimes that I think I'll just leave home.

Client 16, a teenager: I know someone who is selling drugs, and I don't know what to do about it.

Client 17, a single, middle-aged woman: I'm so depressed. I went to the doctor and found out that I'm pregnant. I feel like this is the end of the world.

Client 18, a foreign student: This is my first year in the United States. I had a fellowship during this past year, and applied for one for next year but didn't get it. So, now I am forced to consider a job. You know I am a foreigner and have only a student visa. I am not supposed to work here, and as I don't have a degree from the U.S. it will be very difficult to get a job. What would I do if I couldn't get a job? What kind of life do I expect to lead? How can I go back home without completing my studies? I don't even have enough money to go back.

Client 19, a teenager: Both my parents think I'm still a little kid. I just turned fifteen, and they won't let me go out on a single date with my boyfriend. All of my friends are allowed to date, so I don't see why I can't.

Client 20, a mother of two young children: My husband had a stroke. He'll be in the hospital a long time, and I need money to support us while he's there.

SOLO EXERCISES. Practice all five role communication skills in the triad. Try to have your minisessions run for 15 minutes, and after each session, review the counselor's responses. Make sure each counselor-in-training has had a chance to practice and can consistently give higher level responses.

‖ SUMMARY

‖ This chapter has required your active participation to develop an understanding of the attending, the clarifying, and the supporting role communication skills. After learning about these primary skills and practicing them in the simulated exercises, you should be able to describe these skills, discriminate among higher and lower level responses, and demonstrate appropriate use of these skills in your counseling sessions.

Attending has been defined as the process of actively listening to another person and reporting back to that person what you believe was communicated to you. Attending responses include simple minimal verbals, reflections of content and feelings, accents, paraphrases, and nonverbal signals. High level responses are characterized by demonstrating interest in the other person, maintaining appropriate body language, responding to the underlying feeling and content of the other person, and employing appropriate tonal qualities in your voice.

Clarifying has been defined as the process of clearing up any confusion that may be present in the counseling process. Clarification responses include perception checking, clarification among alternatives, and requests for elaboration. High level responses focus on critical issues or situations that need clarification and demonstrate a sincere and genuine interest in the client.

Supportive responses are designed to communicate your belief and faith in the client, provide emotional security, reduce client anxiety, and provide encouragement. Supportive responses include person-of-value responses, approval responses, consolation responses, and relaxation responses. High level responses show concern for clients, reveal an understanding of the problems that human beings face, and are presented in a warm and relaxed tone.

In addition to reading the chapter and practicing the primary role communication exercises, these skills should be practiced in your daily conversations with other people. These primary skills are extremely useful in all human conversations, and their effective use can improve your understanding of another person's point of view. Because these skills are necessary in each phase or stage of the counseling process, it is crucial that you master them as soon as possible.

REFERENCES AND SUGGESTED READINGS

Benjamin, A. (1987). *The helping interview* (4th ed.). Boston: Houghton Mifflin.

Brammer, L. M. (1988). *The helping relationship* (4th ed.). Englewood Cliffs, NJ: Prentice-Hall.

Cormier, W. H., & Cormier, L. S. (1991). *Interviewing strategies for helpers: Fundamental skills and cognitive behavioral interventions* (3rd ed.). Pacific Grove, CA: Brooks/Cole.

Egan, G. (1990). *The skilled helper: A systematic approach to effective helping* (4th ed.). Pacific Grove, CA: Brooks/Cole.

Evans, D. R., Hearn, M. T., Uhlemann, M. R., & Ivey, A. E. (1989). *Essential interviewing: A programmed approach to effective communication* (3rd ed.). Pacific Grove, CA: Brooks/Cole.

Hackney, H., & Cormier, L. S. (1988). *Counseling strategies and interventions* (3rd ed.). Englewood Cliffs, NJ: Prentice-Hall.

Hutchins, D. E., & Cole, C. G. (1986). *Helping relationships and strategies.* Pacific Grove, CA: Brooks/Cole.

Ivey, A. E., Ivey, M. B., & Simek-Downing, L. (1987). *Counseling and psychotherapy* (2nd ed.). Englewood Cliffs, NJ: Prentice-Hall.

Johnson, D. W. (1990). *Reaching out: Interpersonal effectiveness and self-actualization* (4th ed.). Englewood Cliffs, NJ: Prentice-Hall.

Okun, B. F. (1987). *Effective helping: Interviewing and counseling techniques* (3rd ed.). Pacific Grove, CA: Brooks/Cole.

8

Intermediate Role

Communication Skills

INTRODUCTION

This chapter is devoted to a discussion of intermediate role communication skills. Informing and probing roles are described and illustrated, examples of the effective use of these roles are presented, and exercises are furnished to enable you to practice these skills. A discussion of the counselor's use of silence is incorporated in this section. After careful review of the material in this chapter you should be able to:

- Outline the objectives of the informing and probing roles.
- Describe the purposes and the meaning of silence in a helping relationship.
- Discuss the characteristics that distinguish high level responses from low level responses in informing and probing roles.
- Identify the types of responses used in each of these roles, and cite examples of good and poor uses of these responses in helping interviews.
- Apply these communication skills in appropriate situations.

THE INFORMING OR DESCRIBING ROLE

The informing or describing role is used when the couselor wants to provide the client with specific, concrete, relevant, factual information; or with descriptions or explanations of how various structures work, or how they may be organized. Giving information and describing how entities function or work are essential counselor roles. Providing clients with information is generally done for one of the following reasons. First, counselors need to apprise clients of how the counseling process works, including information about procedural matters, the roles of the client and counselor, what the process is like, what the ground rules are, and what generally can be expected. Second, counselors need to supply clients with specific information necessary for the counseling process and at levels that clients can understand. This factual material is often educational or occupational information, the meaning of test data, and descriptions of appropriate resource materials. Third, counselors frequently need to supply clients with basic information about essential psychological principles, such as the importance of human drives and needs, the themes of different developmental periods of life, the role of anxiety in life, the difficulties of choosing between alternative courses of action, or the inappropriate use of certain coping strategies. Finally, there is the need to inform clients about particular intervention strategies.

Presenting information is not the same as giving advice, suggestions, or directives; it is not value-laden material, but rather objective and accurate factual material about people, places, or things. It should be stated in an impersonal, matter-of-fact, neutral tone (Benjamin, 1987). The kind of information that you provide to a client should be directly related to his or her needs and concerns at any given moment in the counseling relationship. This type of response is employed when the client is not likely to have the information or when the client has some misinformation and it is important for him or her to know the correct information for the counseling process to advance.

The amount of time spent providing factual information varies according to the needs of the client and the type of counseling that is being provided. In some cases a considerable part of an interview or even several interviews may be spent presenting and reviewing important constructs or databases, and in other cases only a minimal amount of time is spent using this role. In either case, you must have solid knowledge and a good information base to use this role efficaciously.

Types of Information or Describing Responses

The information role is typically employed for one of the following purposes:

1. *To structure the counseling process.* Responses of this sort inform the client about how the counseling process is structured. As the counselor, you may need to describe how the counseling process works and the role and responsibilities of the counselor and the client; provide the client with practical detailed information such as time, location, duration, and cost for each session; or inform the client about taping requirements, record keeping, and how confidentiality will be maintained. Providing clients with these descriptions can reduce anxiety and enable clients to verify their expectations (Hackney & Cormier, 1988).

2. *To provide clients with relevant factual information.* This type of response is designed to present clients with important facts that are relevant and necessary to advance the counseling process. Frequently this response communicates information that facilitates the client's self-knowledge and the client's knowledge of the educational and vocational process. This may involve furnishing relevant information about the career-development process; the meaning of psychological tests and test scores; the availability of educational and vocational

resource materials; or specific data about the rules and regulations of a particular educational institution or community agency. The information presented must be clear and relevant to the client's concerns (Egan, 1990).

3. *To inform clients about some basic psychological principles.* Responses of this type present clients with information about important principles of human nature including such phenomena as human drives and needs, the developmental process, the decision-making or problem-solving process, and ways to reframe issues in order to analyze them from another perspective. Helping clients with the problem-solving process may involve providing information about new or different alternatives or the possible outcomes of various alternatives, or stating the fact that decisions and choices usually involve some risk and are rarely made with absolute certitude.

4. *To present information that is relevant to a particular counseling intervention strategy.* When a specific intervention strategy is employed, clients may need to be given an explanation of the rationale for the strategy, the meaning of certain terms used in the process (such as raw score, contingency reinforcement, or covert modeling), their role and what they have to do in this process, and the steps that you plan to use as the process unfolds. Specific information about a counseling intervention technique is often given to clients during the decision-making and working stages of the counseling process.

Different Types of Informing and Describing Responses

The following set of examples illustrates how this role is used.

CLIENT 1, A 23-YEAR-OLD WOMAN: I lost my job about seven months ago. I went on unemployment instead of looking for other work. I thought it would be fun, just staying home. Now, all I do is lie in bed or stay in my nightgown all day. My husband's been complaining that I look like a slob, and the house is a mess. I can't seem to get myself together.

> *Structuring the counseling process:* It sounds like you think coming here for counseling can help you get yourself together. It probably can, but first you must understand that the counselor has no magic. Any change will depend on you and the goals you set for yourself.

CLIENT 2, A 19-YEAR-OLD COLLEGE FRESHMAN: I'm not sure what I'd like to major in at college. My dad said you would have the latest information on the job outlook for economists, lawyers, and accountants.

> *Important factual information.* Yes, we do have a number of resources that can supply you with that information. First, there's the *Occupational Outlook Handbook,* and information about each of the fields you mention is in the occupational library files. Also, the respective departments have very current information posted on their bulletin boards.

CLIENT 3, A COLLEGE SOPHOMORE: I've been looking over the results of the Strong Interest Inventory that you returned to us last week, and I don't understand them. It seems to indicate that I should become a doctor or a dentist, and I don't want anything to do with those fields.

> *Information about psychometric data:* The Strong Interest Inventory is not an aptitude test. It does not tell you whether or not you have certain skills. What it does say is that your interests are similar or dissimilar to people who are in certain occupations.

CLIENT 4, A HIGH SCHOOL JUNIOR: I'm not sure what I want to do with my life. There are lots of things that appeal to me. I think it would be fun to major in English in college and read and analyze great literature, but then again something exotic like zoology or anthropology might be even more interesting. Maybe you have some tests or something that can tell me what I should do.

> *Information about some basic psychological principles:* Being open to a variety of occupational choices is healthy at your age. Decisions about occupational choices involve learning more about yourself, learning more about the world of work, and engaging in some exploratory activities. Even though there isn't any test that can tell you how to spend your life, some tests might be useful in beginning this exploration process by first helping you learn more about yourself.

CLIENT 5, A COLLEGE JUNIOR: I'm afraid I would blow any job interview. I won't know what to say. . . . My mind goes blank. I have good grades, yet I panic when I have to speak to someone I don't know. You helped me once before when I almost dropped out of college. Can you help me now?

> *Information relevant to a particular intervention strategy:* It sounds like you have a great deal of anxiety when you are faced with a new or an unknown situation. If you want me to help you overcome this fear, you will need to tell me quite a bit about your life, particularly about other events or situations that have caused a similar type of response in you. After we obtain your life history, I will teach you a systematic way of relaxing.

Levels of Informing and Describing Responses

The informing and describing role can be used in ways that differentially affect its potency. Individuals who are effective in using this role provide relevant information, describe things or circumstances accurately, respond in ways that are understood by their clients, and ensure that the material is understood. If the information is complex, it should be broken down into units that the client can assimilate (Cormier & Cormier, 1991; Evans, Hearn, Uhlemann, & Ivey, 1989). Counselors who employ this role at appropriate times advance the counseling process and communicate that they are trained professionals knowledgeable about their field. Information responses are ineffective and counterproductive when unnecessary information is imposed on the client, when the responses fail to provide meaningful and complete descriptions, or when they are given with evaluative overtones or are otherwise value laden. On the four-point scale, the effectiveness in using this role can be rated on factors such as furnishing accurate and relevant information, employing appropriate terms and phrases, providing data and descriptions that can be readily understood, presenting materials when clients need the information and are ready to receive it, and providing information in a value-free manner.

> *Level 1: Poor use of the role.* The counselor uses this role in an ineffective way when he or she describes or explains matters of concern incorrectly, uses distracting gestures, informs the client prematurely, or provides irrelevant material. The role is also counterproductive when it is used excessively or in a tone that conveys a belaboring or demeaning manner.
>
> *Level 2: Mediocre use of the role.* This role is used somewhat ineffectively when the counselor provides the client with partial or incomplete information, when the information presented is correct but not germane to the client at the point it is given in the counseling process, or when the counselor's messages are too wordy or the terminology too complex.
>
> *Level 3: Good use of the role.* The counselor is using this role effectively when he or she responds to the needs of the client by explaining phenomena in sufficient detail to be useful to the client at the time it is given in the counseling process.
>
> *Level 4: Excellent use of the role.* The counselor is most effective when he or she describes things completely and responds to the client's need for information propitiously. The information is provided in a warm tone and with gestures that can help the client relate to the factual material presented.

Different Levels of Informing and Describing Responses

The following illustrations of responses are provided so that you can see the differences between effective and ineffective use of the informing role. Each of the counselor's responses has been rated using the Level-of-Response Scale, and the reasoning behind each of these ratings is indicated. Study these examples to determine if you agree with the ratings. You may find that you do or do not agree; any response may be rated differently depending on the paraverbals that the reader hears. Thus, it is important to discuss these ratings and the reasons behind them with your classmates and your supervisor so that any differences can be understood.

CLIENT 1, A 22-YEAR-OLD: About three years ago I got stuck in an elevator. I was alone, and I was very scared. But after about five minutes the elevator started up, and I was fine. Since then, I can't ride in an elevator without feeling boxed in, shaking, and getting sweaty palms. It's okay when there are stairs and I can walk up to where I want to go, but sometimes I avoid visiting friends because they live above the sixth floor.

Responses	*Ratings and Reasons*
1. Fears are sometimes hard to overcome, but with a commitment to change and some hard work, fears can be overcome or at least lessened so they can be handled.	2.4 The counselor briefly outlines the counseling process. Although the words are accurate, it sounds like a lecture.
2. I think we can help you ride in an elevator again. First, we will need to discuss any other fears that you may have.	2.5 This initial outline of a counseling strategy is good but quite limited.
3. Your fear of elevators can be overcome. You must learn to get control of your emotions.	1.0 This authoritarian response is not informative or descriptive.
4. I sense a feeling of helplessness on your part, and I will be glad to work with you. It will take us some time to reduce your anxiety, but I believe we can have you visiting your friends again.	3.0 This very brief description of the counseling process shows an understanding and feeling for the client's problem.

CLIENT 2, A COLLEGE SOPHOMORE: My parents want to know what I plan to do after college. They think I should major in accounting or economics or something that has a definite orientation toward an occupation. I want to major in history, which they believe is a total waste of time. Do you have any tests that I can take to show them that accounting is not for me?

Responses	*Ratings and Reasons*
1. I know you are feeling a great deal of pressure, but I'm afraid that in your present condition we would not get an accurate picture from any test.	3.0 This information about tests is appropriate and helpful.
2. Parents always want to live your life for you. Yes, I have a good inventory that will reveal your true interest patterns.	1.0 This attempt to identify with the client is poor, and the information about testing is probably misleading given the client's motivation.
3. You don't want to do something just because your parents want	2.7 Shows awareness of client's feelings and situation.

you to. We have some tests that may help; however, they may or may not come out the way you want them to.

The information about testing, although brief, should further the client's awareness.

4. You feel pulled in two directions, your parents and your own, and you'd like to resolve this in some way. I feel the pressure you have, but I do not believe a battery of tests will ease this pressure. It may prove that you are right, or it may prove that your parents are right about your abilities. But I think it will prove neither.

3.5 The counselor shows his perception of the client's struggle and provides information about tests that should be beneficial to the client.

CLIENT 3, A MIDDLE-AGED WOMAN: I have to do something. My children are all grown. My husband is extremely busy with his work. I would like to get a job, but I don't know how to begin.

Responses

1. Many women return to the labor force when their children are grown. Our office has helped many women returning to work. Working together we should be able to help you find a job that will meet your needs.

Ratings and Reasons

2.8 The counselor presents factual information and begins to structure the counseling process.

2. You really have an empty feeling and sense you must do something. I think we can work with you on this problem, and we can help you begin by talking about you as a person—your likes, dislikes, and previous accomplishments.

3.0 The counselor shows awareness of the client's concern and briefly outlines the initial steps of the counseling process.

3. Yes, well, the first step is to take a battery of tests that will outline your abilities and competencies. Then we will match these up with various occupations and job openings, and see which one is the best for you.

1.5 The counselor falsely implies that vocational selection is primarily a psychometric function.

4. There are several steps. These involve helping you understand yourself a little better, learning about the kinds of work available, and exploring the occupational choices that appear to meet your needs.

2.8 The counselor briefly outlines the vocational counseling process.

Practice Informing and Describing Exercises: "How Do I Provide Specific Information or Descriptions of How Things Function?"

The following exercises have been prepared to assist you in developing your informing role skills. The discrimination exercises afford you the chance to further practice identifying clients' underlying messages and the different ways to inform and describe various things to clients, and to learn to discriminate between more and less effective responses. The communication exercises enable you to practice giving high level informing and describing responses in a variety of client situations. Finally, the solo exercises give you the chance to enhance your attending, clarifying, supporting, and informing role communication skills.

DISCRIMINATION EXERCISES. For each of the following excerpts, first identify the feeling and the content underlying the client's statement, and then specify the type of informing and describing response illustrated; finally, indicate the appropriateness of the response by rating the level of the response. Please be prepared to share the reasons for your ratings with your classmates and your supervisor. You may find that one or more of your classmates read the excerpts with another tonal quality than you did and consequently rated the response quite differently.

CLIENT 1, A RECENT HIGH SCHOOL GRADUATE: I never seem to be a success at anything. Everything I try seems to end in failure. I've never had a job that lasted for more than six months.

Underlying message: _____

Responses

Type and Level of Response

1. We should plan to meet at this time for an hour each week to explore your experiences and help you rediscover your strengths.

1. _____

2. Taking one or two of the career inventories can help us under-

2. _____

stand your occupational interests. They can be done in an hour and ask questions about the kinds of things you enjoy doing.

3. You are facing up to the difficulties of finding the right job. Your experiences may have helped you become more aware of yourself and your vocational interests.

3. _____

4. You were successful in graduating from high school. Many people are not. It's not uncommon for people to experiment with different jobs before they find one suitable to their own interests.

4. _____

CLIENT 2, A HIGH SCHOOL FRESHMAN: I get really worried about my friends—sometimes I can't get to sleep at night I'm so worried. They all recently began to use the real heavy stuff.

Underlying message: _____

Responses

Type and Level of Response

1. You have good reason to worry about your friends. The stuff they are using is dangerous and destructive.

1. _____

2. It sounds like you're in a tough spot. You've taken the first step to deal with this concern. We will need to plan some further steps.

2. _____

3. I understand how upset you are. I have no magic to solve this problem, but if you are willing to work with me, we can resolve this issue.

3. _____

4. And you hope I can tell you what to do about it. I wish I could just tell you the answer. That would make it easy, but that's not the way counseling works. If we work together, you can discover how to tell yourself what to do.

4. _____

CLIENT 3, A COLLEGE SOPHOMORE: My father's giving me a hassle because my grades aren't that great, and he keeps telling me to make up my mind about my career. I get sick of him telling me that all the financial sacrifices he's making better be worth it. Oh, maybe I shouldn't be here in the first place . . . maybe I ought to be out working.

Underlying message: _____

Responses	*Type and Level of Response*
1. It sounds like if you don't improve your grades, you will have no choice but to be out working. We need to look seriously at what is bothering you so you can free yourself and be able to concentrate more on your studies.	1. _____
2. The pressure coming from your father is only part of the problem we are dealing with. How you handle this type of pressure is another part.	2. _____
3. It sounds like you want to work at changing things. Counseling can help you clarify what's going on and probably help you make a decision. It means an investment of time for both of us.	3. _____
4. I'm happy you can ask yourself the questions "Should I be here? Maybe I ought to be out working?" There are several interest inventories that are easy to take. They can help you make a decision about where you're at in this whole education scene.	4. _____

CLIENT 4, A 25-YEAR-OLD WOMAN: I've been constantly tired lately. I just can't seem to keep my mind on anything. I have so much work to do, but I just can't get up the energy to do it.

Underlying message: _____

Responses	*Type and Level of Response*
1. There are several reasons why you may have this feeling. You	1. _____

may have accepted too many
responsibilities or set a demanding pace for yourself. You may
be fighting something and need
some vitamins, a good rest, or
both. But in your case, your
lack of energy and inability to
concentrate appear to have
other causes.

2. You must stop thinking the   2. _____
   way you do. You keep saying
   everything must be done
   perfectly.

3. Sometimes the counseling pro-   3. _____
   cess helps us get at the bottom
   of why you are feeling the way
   you do. We can explore some of
   the things that are going on in
   your life, and I feel confident
   you will discover some new
   ways to look at things.

4. It sounds like you are confused 4. _____
   about what's going on inside
   you. There are some simple
   tests you can take that save time
   and energy and help me to
   focus more quickly on your
   difficulty.

CLIENT 5, A 30-YEAR-OLD ENGINEER: I really don't know what I should
do. My boss is driving me crazy. He knows that I'm scared to death about flying. And now he wants me to go to Saint Louis to estimate a job.

Underlying message: _____

Responses   *Type and Level of Response*

1. We need to separate the pres-   1. _____
   sures of your job from your fear
   of flying. Through desensitization you might be able to
   overcome your fear; however,
   your relationship with your
   boss may be quite a different
   issue.

2. It sounds as though you feel   2. _____
   that I can tell you what to do. I
   can do that, but it will be much
   more beneficial to look at some
   different ways to respond to
   your boss and their
   consequences.

3. By exploring some of the feel-   3. _____
   ings you have about the situa-
   tion, you probably will discover
   what you should do. However,
   if you need an immediate
   response to your boss, we can
   talk about it today and help you
   decide on a way to approach
   this problem.

4. You seem to be torn between   4. _____
   your desire to succeed and your
   fear of flying. There are some
   things that you can do to
   reduce your fear of flying to
   manageable proportions. But it
   will take practice to learn, and
   you must be willing to try.

CLIENT 6, A 65-YEAR-OLD WIDOW: My children don't understand. They
question me about everything—where I go, what I do, who I'm dating. They
think I've flipped.

Underlying message: _____

Responses   *Type and Level of Response*

1. As people grow and change, it   1. _____
   is not uncommon for children
   to assume parental roles concerning their parents, and you
   are experiencing this now. It is
   quite a role reversal.

2. Your children's perception of   2. _____
   you and your own perceptions
   are quite different. You don't
   want to live within the mold
   they have created for you. The
   counseling process can help you
   learn how to tell your children
   about your own needs.

3. During the counseling process
we will explore some of the
feelings you have about your
children as well as the feelings
you have about yourself. By exploring these feelings, we
should have a better understanding of the situation.

3. _____

4. We will be able to meet here
every Thursday at 4:00 p.m.
The fee will be _____ dollars.
If you need to cancel an
appointment, please call as early
as possible.

4. _____

CLIENT 7, A 21-YEAR-OLD MAN: I'm so confused. Here I am engaged to
a wonderful girl. Everything is going my way, and yet I'm not content with myself.
I feel as if there's something missing in my life.

Underlying message: _____

Responses

Type and Level of Response

1. The counseling process can
help explore the feelings and
expectations you have in your
life, and that can be helpful in
discovering the basis of some of
the confusion you feel. How you
change things will be up to you.

1. _____

2. Sometimes people can't get at
the root of their confusion.
Counseling can help. If you are
interested in going that route, I
can set up an appointment for
you next week.

2. _____

3. Your test results seem to indicate that you are very creative,
yet the job you are working at
gives you no room for creativity.
They also indicate that you like
to work with people, and right
now you are tied to a machine. Part
of your difficulty may be that
most of the day, five days a
week, you are doing a job
you hate.

3. _____

4. Yes, and you keep telling your-   4. _____
    self that you're no good. This
    negative self-talk is causing you
    a lot of pain.

CLIENT 8, A 30-YEAR-OLD MOTHER: I can't cope with my children anymore. They come home from school, turn on the radio, tie up the phone, and totally ignore me when I speak to them. We fight all day long, and my nerves feel tied up in knots.

Underlying message: _____

Responses   *Type and Level of Response*

1. We will need to look at some of   1. _____
    your parenting practices. I think
    you may be reinforcing your
    children's poor behavior
    unknowingly.

2. The fact that your children may   2. _____
    be healthy adolescents is not
    consoling. Counseling can help
    you become more comfortable
    with yourself and your role as
    mother, and show you how to
    communicate more effectively
    with your children.

3. Sometimes, talking about these   3. _____
    things helps clarify exactly what
    it is you are feeling. Then it be-
    comes easier to deal with the sit-
    uation. That's what counseling
    involves: talking, changing some
    things, and learning to cope better.

4. I understand what you're saying.   4. _____
    As we begin to talk about you
    and your family, I will want to
    review with you some basic
    psychological needs that all
    children have and that many
    children try to satisfy in in-
    appropriate ways.

CLIENT 9, A MIDDLE-AGED MAN: Mary and I have been married for twenty years. It was really good in the beginning, but these past few years our marriage has been terrible. I'm considering a divorce. I just can't take it anymore, it's not worth it.

Underlying message: _____

Responses	*Type and Level of Response*
1. It sounds like your dream has reemerged. Individuals your age often assess where they are and compare it to where they want to be. It's one of the most difficult times in a marriage. Earlier you were busy establishing yourself in your job and your wife was busy with all the homemaking chores—your relationship never grew. Now it's obvious that you've lost contact, and you feel you are living with a stranger.	1. _____
2. When there are difficulties in the marriage, very often both parties are struggling with communication. Some blaming is going on on both sides. Walls build up and get higher and thicker. Before that happens, we can look at the whole picture as both of you see it. So often it is the system in a family that's in trouble and no one person is to blame.	2. _____
3. Through the counseling process, you will be able to discover more clearly what is going on with you and Mary. The decision for or against divorce might then be made with more understanding.	3. _____
4. Since the problem seems to be between you and Mary, it would be a good idea for both of you to come to talk together. The situation works out better if both of you are involved in the counseling process.	4. _____

CLIENT 10, A 28-YEAR-OLD WOMAN: I'm depressed all the time. Everything I do turns out to be a failure. My family criticizes me. My boss yells at me. Nothing is right for me. Please help me.

Underlying message: _____

Responses   *Type and Level of Response*

1. I'm glad you came. You are cer-   1. _____
   tainly confused about what is
   going on inside you, and you
   need some help. We will talk for
   a while and see if we can learn
   more about you and how we
   can work together.

2. You are not a failure at every-   2. _____
   thing. You dress attractively;
   you take care of your physical
   self and your appearance. Your
   depression has specific causes
   and is not caused by everything
   in your life, although you may
   feel like it is at the moment.

3. Depression has a way of feeding   3. _____
   itself, and often we make more
   mistakes in activities or in interpreting other responses because
   of how we feel. This self-
   defeating behavior makes us
   more depressed and angry with
   ourselves.

4. In order to help you, we will   4. _____
   have to work together. It means
   a process of counseling that involves hard work for us, but if
   we meet consistently every
   week for an hour, I'm sure
   things will get better for you.

COMMUNICATION EXERCISES. For each of the following client statements, please formulate at least two high level informing or describing counselor responses.

> *Client 11, a young woman, 17 years old:* It's my father. For years now, I can't remember the last time he came home sober. He's drunk all the time now. He thinks that he's fooling everybody, but he isn't. All my friends know what's going on. I can't take it anymore.
>
> *Client 12, a 25-year-old woman:* I still can't believe I'm divorced. I don't have anyone to share my life with now. And I have to start the dating scene all over again. I don't know what I'm going to do!

Client 13, a 40-year-old woman: My problem is that I have twin boys age seven, and I can't find the energy or patience to be with them anymore. I wish I were back at work—they're driving me crazy. I feel so guilty about this.

Client 14, a high school senior: I think I'm going to quit school. I can't stand the courses. I don't feel I'm learning anything.

Client 15, a 50-year-old man: I caught my daughter smoking marijuana, and I've put her on restriction. It really hurt me, and now I'm wondering what else she might be doing.

Client 16, a 23-year-old man: I'm thinking that I've made the biggest mistake of my life. I got married two months ago, and now it's not working out. We live upstairs from her parents. My wife and I fight constantly, and her family always takes her side.

Client 17, a 23-year-old homemaker: My drinking is starting to frighten me. It seems like I'm climbing into the bottle more and more often.

Client 18, a 22-year-old college student: In two months I graduate from college. I don't have a job to go to or any plans, for that matter. I don't really want to move back home, but there's nothing else for me to do. I just don't know.

Client 19, a high school freshman: I feel very lonely. I don't seem to have any friends. When lunchtime comes I eat by myself. I don't see anyone after school.

Client 20, a 21-year-old man: I'm twenty-one years old and am basically dissatisfied with my life. I'm not happy with my job—I'd like to go back to school, but I don't know for what. I'm wasting my time, and I see my whole life as meaningless.

SOLO EXERCISES.   Team up with two of your classmates to form a triad. Take turns being "the counselor," "the client," and "the supervisor." Role-play in order to further develop your skills. Use the four role communication skills you have learned, and try to employ a variety of responses within each role. Spend at least 10 minutes in each minisession.

THE INQUIRING OR PROBING ROLE

The inquiring or probing role is often employed when you as the counselor seek to obtain further information from the client. It may be used when you want to begin an interview; encourage the client to speak or elaborate on a topic; help a client focus or describe his or her thoughts, feelings, and behavior more completely; assist the client in identifying concrete examples of his or her concerns; or discover what resources the client has at his or her disposal. The inquiring or probing role can be employed to guide the discussion and help the client obtain certain insights. This role is most often exemplified by interrogative statements or questions, but it also includes incomplete sentences, in which the client is expected to complete the thought; accents, in which the counselor repeats part of a client's statement to elicit further information; and paraverbal or nonverbal messages, in which the counselor's tonal quality or a quizzical look conveys the probe in a clear manner.

Inquiries or probes should normally be made in the form of open-ended questions or statements. This type of questioning requires more than a simple yes or no response. Open-ended questions encourage clients to share their feelings and thoughts with the counselor and permit clients to respond in the way they prefer. Open-ended questions often contain the words *how, what, when, where,* or *who.* These words are associated with seeking out different kinds of information: *how* probes into procedures or processes that caused or preceded the stated event, thought, or feeling; *what* inquires about specific facts and details; *when* and *where* look for information concerning circumstance and occasion; and *who* seeks information about individuals (Evans, Hearn, Uhlemann, & Ivey, 1989; Ivey, Ivey, & Simek-Downing, 1987).

"Why" questions seek reasons for thoughts, feelings, or behavior and should generally be avoided because they ask clients to justify their actions or speculate about their motives when they may be unaware of the causes of their feelings and behaviors (Brammer, 1988). This type of question often provokes defensive feelings (Evans, Hearn, Uhlemann, & Ivey, 1989). "Why" questions can also indicate approval or criticism or offer advice to a client that is counterproductive to the inquiring role (Johnson, 1990).

Probes or inquiries can be phrased in a statement form rather than a questioning form when the counselor wants to avoid creating an interrogative atmosphere. For example, "What happened before you did that?" can be restated as "Describe the events that led up to that situation." Benjamin (1987) refers to this as the indirect question, and Okun (1987) maintains that it is highly preferable to use the statement form of the probe. Many counselors-in-training have found this approach more effective than the direct question. When they used it, their clients gave more elaborate responses. The statement form acts as an encouragement for clients, and it appears to help them elaborate and continue to express themselves.

When you need to find out a specific or concrete detail, a close-ended question is appropriate; otherwise, it is usually better to avoid using this type of probe. Close-ended probes typically can be answered with a very brief retort, or a yes or no response. They tend to elicit factual materials that often have little bearing on the major concerns of the client and frequently do more to satisfy the counselor's curiosity than to advance the counseling process. Close-ended questions often contain a form of the verbs *to be* or *to do,* such as "Are you feeling okay?" or "Do you like that?"

You should minimize the use of the probing role in order to avoid creating an interrogative atmosphere. This climate allows your client to play a more passive role in the counseling process and contributes to dependence rather than independence on the client's part. Counselors who probe a great deal communicate that they are willing to take the major responsibility for the content and direction of the counseling sessions. Furthermore, the excessive use of the role can put your client on the defensive or communicate disapproval of the client's activities.

Types of Probing and Inquiring Responses

The following examples illustrate the use of various types of inquiring and probing responses.

1. *Open-ended questions.* Open-ended questions are often used to:
   a. open the interview—for example, "What would you like to talk about today?"
   b. invite the client to speak or to elaborate on a topic—for example, "Could you tell me more about that?"
   c. ask the client to focus or explore a feeling or thought more completely—for example, "How did you feel then?"
   d. request an example or an illustration of a situation—for example, "Would you give me an example of that?"
   e. help a client identify internal or external resources—for example, "Do you think you can handle that situation?" or "Who can give you some information about that?"

   The question must always be relevant to the concern of the client and should encourage the client to discuss any thoughts, feelings, or behaviors that would facilitate the counseling process.

2. *Exploratory or elaborative statements.* Instead of asking open-ended questions, many counselors prefer to use statements that ask clients to elaborate or explain a situation, thought, or feeling more completely. Notice how the five following examples inquire into the same areas as the open-ended questions illustrated above:
   a. open the interview—for example, "Tell me how I can help you today" or "I wonder how I can help you today"
   b. invite the client to speak or elaborate on a topic—for example, "Describe that situation more completely" or "I'd be interested in hearing more about that situation"
   c. ask the client to focus or explore a feeling or thought more completely—for example, "Explain how you felt when that happened to you" or "I'm sure you had some strong reactions to that situation"
   d. request an example or an illustration of a situation—for example, "Describe a typical incident when that happens to you" or "I'd like to hear about a particular time when that occurred"
   e. help a client identify internal or external resources—for example, "Tell me how you could handle a situation like that" or "I wonder if you know two or three people who could give you some specific information about that"

   In each of these examples two different styles of the statement form of the probe have been used. One form is more direct, whereas the other is more indirect. The open-ended statement form of the probe normally facilitates a more comfortable climate and often leads to richer and more complete responses than open-ended questions.

3. *Close-ended questions*. It is often necessary to inquire about relevant details and to obtain specific information. These inquiries are designed to help complete the picture that the client is painting about his or her life. This information is usually obtained by close-ended questions such as:
   a. "How old were you then?"
   b. "Do your parents know about this?" and
   c. "Did you have to see a doctor?"
   Multiple use of this type of questioning should be avoided.

Different Types of Probing or Inquiring Responses

The following examples illustrate the use of the inquiring response. Compare the purposes and effectiveness of the open-ended questions used in the first example, the interrogative statements used in the second example, and the close-ended questions used in the third illustration. Also note how many of these examples use the attending or supporting role before the probe. This combination is used to convey to the client that he or she is being listened to but further information is needed to advance the counseling process.

CLIENT 1, AN 18-YEAR-OLD WOMAN: My mother is always criticizing me for the way I look and dress. She thinks there's something wrong with me because I don't date. It's not that I'm not interested, but most men I've met so far are dull and boring.

1. *Interview opening:* It sounds like an awful situation. What is the major issue that you would like to talk about today?
2. *Elaborative inquiry:* You're very upset. I'm not sure whether it's because of your mom, your social life, or both. Could you tell me more about yourself?
3. *Exploration of feelings:* It really gets you down when someone picks on you. It's doubly hard when things aren't going the way you would like them to. How do you feel when your mother picks on you?
4. *Request for an example:* I know how you feel when people don't see things the way you do. Would you give me an example of how your mother criticizes you?
5. *Inquiry about resources:* It's painful to be picked on, particularly by someone close to you. How do you meet members of the opposite sex?

CLIENT 2, A HIGH SCHOOL SOPHOMORE: My parents don't understand. Everything I do seems wrong to them. It's impossible living with them.

1. *Interview opening:* It sounds like you want to talk about your parents today. Tell me about them.
2. *Elaborative inquiry:* Your parents seem to you to be supercritical. You can't do anything right, and it's tough when you're home. I'd like to hear more about your situation.

3. *Exploration of feelings:* It hurts a lot when you're treated like a child. Adults, and your folks in particular, want you to do things their way. Tell me how you feel about that.

4. *Request for an example:* It's tough to live with people who don't understand you. Give me an example of what they say to you and how you respond.

5. *Inquiry about resources:* It is painful to be misunderstood. Name someone—a friend, anyone else in your family—that you can talk to who understands you.

CLIENT 3, A 32-YEAR-OLD MAN: My wife just had a hysterectomy. She'll never be able to have children. She's only been out of the hospital for about two weeks, and now all she'll talk about is adopting a child. Before this, we'd always thought we'd wait a couple of years before starting a family. I don't know what to say to her.

1. *Interview opening:* It sounds like you need to learn how to improve your communication with your wife. Would you like to learn how to do that today?

2. *Elaborative inquiry:* You sound like you're a little stunned. You're concerned about your wife and her present health, and she's already planning for the future. How is your relationship with your wife?

3. *Exploration of feelings:* She shifted gears and you didn't. Wow! Did you feel surprised or angry about this?

4. *Request for an example:* You're startled and confused. Your wife is still recuperating, and she changed her mind and expects you to change your mind too—not only about the time to start a family, but about this whole idea of adoption. Did you ask her what this would mean to your relationship?

5. *Inquiry about resources:* She wants a family right now, and you're not sure about that. Do you know how to go about adopting a child?

Levels of Probing and Inquiring Responses

The inquiring role can be quite effective when the client is given the time to respond to the probe and one question is not followed by another (Cormier & Cormier, 1991). The skillful use of the probing role is a function of the relevancy of the inquiry; the appropriateness of the depth of the probe; the pacing and timing of the questions; and the warmth, respect, and genuineness associated with the probe. The role is very ineffective when double or multiple questions are used. This "ping-pong" effect (Hackney & Cormier, 1988) can cause confusion or create an atmosphere of cross-examination (Evans, Hearn, Uhlemann, & Ivey, 1989). Asking too many questions in any given session also tends to reduce the client's personal responsibility for the counseling process, increase client dependency, and encourage socially acceptable responses (Brammer,

1988). Seeking tangential or superficial material, satisfying the counselor's curiosity, obtaining excessive factual data, or using a brusque manner can also create an interrogative atmosphere, which is counterproductive to the counseling relationship.

The inquiring role is used throughout the counseling process; however, it is usually not employed extensively in the initial stage of counseling because it intimates that counseling is a question-and-answer or cross-examination process. The probing and inquiring role can be used at various levels of effectiveness and rated on this four-point scale:

Level 1: Poor use of the role. The counselor is using this role ineffectively when he or she probes or inquires about irrelevant data; uses antagonistic or accusative questions such as "Why did you do *THAT?*"; probes to satisfy his or her own curiosity; uses a voice quality or body movement that communicates an abrupt, authoritative, or dull uninterested attitude.

Level 2: Mediocre use of the role. The counselor is somewhat ineffective when he or she is probing for related but unessential information; when the inquiries are poorly phrased; and when too many questions cause an interrogative atmosphere.

Level 3: Good use of the role. The counselor is effective in using this role when his or her probes are relevant and well phrased; when the role is used at a suitable time in the counseling process; and when the counselor's voice quality shows interest and concern for the client.

Level 4: Excellent use of the role. The counselor is most effective when his or her inquiries are phrased in a way that encourages the client to go beyond the surface or stated message. Normally, this role is most effective when it is used sparingly in the counseling process.

Different Levels of Probing or Inquiring Responses

The following examples are provided so that you can learn to distinguish among the levels of response. Similar to the previous exercises, three client statements are presented and each statement is followed by four different probing responses. The responses are rated according to the Level-of-Response Scale, and the reasons for each of these ratings are indicated. Study these examples to see if you agree with the ratings and the reasons stated. You may find that you do not agree with these ratings. Because the nonverbal and paraverbal aspects of the client's statements and the counselor's responses cannot be adequately conveyed by the printed word, it is quite possible that you will read one or more of these statements differently from the person who rated them. Any counselor response may be rated quite differently depending on how it is heard. It is extremely important that you state the reasons for your rating so that you can discuss any differences in class with your supervisor and classmates.

CLIENT 1, A MIDDLE-AGED WOMAN: My twenty-year-old daughter has been giving me problems. She stays out all night and does God knows what. When I ask her what she did, she tells me that it's none of my business.

Underlying message: _____

Responses	*Ratings and Reasons*	
1. What do you believe about your daughter?	2.3	Open-ended question requesting the client to elaborate, but the focus is placed on the daughter and not on the mother.
2. You're frightened and worried about your daughter. You feel your relationship with her is not what it should be. Tell me more about your relationship with her.	3.0	Shows awareness of client's feelings. Uses the direct statement form of probe.
3. Your daughter is irritating you. She doesn't want to live by your rules. I'm wondering what goes on inside of you when she says it's none of your business.	3.5	Counselor reflects client's content and feeling. The indirect statement form of the probe focuses on the client's feelings.
4. Any relationship you had with your daughter has been shattered, and you feel very badly about it. Tell me when you first noticed that you were having problems.	2.5	The counselor is aware of the client's concern; however, the close-ended question probes into material that may not be appropriate at this time.

CLIENT 2, A 29-YEAR-OLD MAN: Every time I go into a singles' bar, I feel lost. If I try to talk to a girl, I stutter and stammer. If I try to make small talk, I get embarrassed. I wish I was more self-confident.

Underlying message: _____

Responses	*Ratings and Reasons*	
1. You feel pretty awful when you go into a bar. Tell me about other situations or other times when you have this same feeling.	3.5	The counselor is showing his concern and then requests appropriate information using the declarative form of the probe.
2. You'd like to feel more self-confident, not only in bars but	2.5	This response demonstrates the the counselor's awareness, but

in dealing with girls in general. Is this lack of self-confidence evident in other aspects of your life?

the close-ended probe is likely to result in a yes or no response.

3. You probably feel quite confident in other dimensions of your life. Tell me about them.

2.0   Assumes too much about the client. Open-ended probe would be good if first statement was appropriate.

4. You wish you were more self-confident when you're in a singles' bar. What do you do when you get embarrassed?

2.3   Shows awareness of client's problem, but the close-ended probe is inappropriate.

CLIENT 3, A 22-YEAR-OLD MAN: My mother always bossed me around when I was growing up. Now, every time my wife tells me what to do, I see red. I start yelling and screaming and telling her to mind her own business. Last night, she said she would leave me if I didn't change.

Underlying message: _____

Responses

1. You can't stand being bossed around. Describe a typical situation when you lose your temper.

2. When someone tells you what to do, you come out fighting. That reaction sure doesn't help your marriage. How is your communication with your wife at other times?

3. You're upset and angry. You want to be in charge of yourself, but you can't. Tell me about other situations in your life where you do feel in control.

4. I understand. Tell me how you feel when your wife tells you what to do, and how you react when she or others boss you around.

Ratings and Reasons

3.0   The counselor shows her awareness of the client's concern. Request for example appears germane.

2.5   Good initial response to client's concern. However, the close-ended question will probably lead to a limited response.

3.0   The counselor's response shows her understanding of the problem. The open-ended probe seeks to discover if the problem is only in this relationship.

2.5   The counselor's request for the client's feelings and the request for the example are too much to ask at one time.

Practice Probing and Inquiring Exercises: "How Do I Say 'Tell Me More About That'?"

These exercises have been provided to help you enhance your probing and inquiring role skills. The discrimination exercises will afford you the chance to obtain additional practice in identifying the client's underlying messages, in recognizing the different ways to probe or inquire for different purposes, and in discriminating between more and less effective responses. The communication exercises will enable you to practice giving high level probing or inquiring responses in a variety of client situations. Finally, the solo exercises will give you the chance to polish your attending, clarifying, informing, and probing role communication skills.

DISCRIMINATION EXERCISES. For each of the following exercises, first, indicate the feeling and the content of the client's message. Second, point out whether the probing and inquiring response is an open-ended question, an open-ended statement, or a close-ended question. Finally, rate the *level* of the response. Be ready to discuss the reasons for your ratings with your classmates and your supervisor. You may find that the excerpts were read in another way by one or more of your classmates; thus, they may have rated the response differently. It is important to understand the reasons for the different ratings.

CLIENT 1, RECENT HIGH SCHOOL GRADUATE: I never seem to be a success at anything. Everything I try seems to end in failure. I've never had a job that lasted for more than 6 months.

Underlying message: _____

Responses	*Type and Level of Response*
1. I'm not sure I understand when you say everything ends in failure. Tell me more about yourself and your job history.	1. _____
2. You are telling me you have no success keeping jobs, and I understand why that upsets you. Tell me about your last job and why you left it.	2. _____
3. It's as if you just can't win at anything you try to do. Did you have a specific career plan in mind when you graduated?	3. _____

4. Tell me about the last job that
   you had, how you got it, and
   what caused you to be unable
   to keep it.

4. _____

CLIENT 2, A HIGH SCHOOL SENIOR: I get really worried about my friends—sometimes I can't get to sleep at night I'm so worried. They all recently began to use the real heavy stuff.

Underlying message: _____

Responses   *Type and Level of Response*

1. When did you first begin to feel   1. _____
   this way?

2. Drugs are scary things. What did   2. _____
   you do when they acted this
   way?

3. How are you affected or in-   3. _____
   fluenced by their actions in other
   ways?

4. Your concern is causing you to   4. _____
   lose sleep. Tell me more about
   your relationship with your
   friends.

CLIENT 3, A COLLEGE SOPHOMORE: My father's giving me a hassle because my grades aren't that great, and he keeps telling me to make up my mind about my career. I get sick of him telling me that all the financial sacrifices he's making better be worth it. Oh, maybe I shouldn't be here in the first place . . . maybe I should be out working.

Underlying message: _____

Responses   *Type and Level of Response*

1. You and your father seem to   1. _____
   disagree about grades and your
   indecision about a career. Tell
   me about your grades and your
   feelings about school.

2. Tell me about any thoughts,   2. _____
   ideas, or dreams you've had
   about choosing a major or occupational field.

3. Who else can you or do you   3. _____
   talk to about your school work
   and your future plans?

4. I understand. I'd be interested in   4. _____
   hearing more about this.

     CLIENT 4, A 25-YEAR-OLD WOMAN: I've been tired constantly lately. I just can't seem to keep my mind on anything. I have so much work to do, but I just can't get up the energy to do it.

Underlying message: _____

Responses   *Type and Level of Response*

1. Has your physician made any   1. _____
   suggestions to you?

2. You have no energy for doing   2. _____
   what you have to do. Tell me
   more about yourself and what
   you expect of yourself.

3. I wonder what a typical day in   3. _____
   your life is like. I want to know
   what you do from the time you
   get up in the morning until the
   time you go to bed at night.

4. Describe any unusual changes or   4. _____
   events that have happened in
   your life recently.

     CLIENT 5, A 30-YEAR-OLD ENGINEER: I really don't know what I should do. My boss is driving me crazy. He knows that I'm scared to death about flying. And now he wants me to go to St. Louis to estimate a job.

Underlying message: _____

Responses   *Type and Level of Response*

1. Your boss asked you to fly. Tell   1. _____
   me about any other things that
   he has asked you to do that you
   feel you cannot do.

2. I'm wondering what it is that is   2. _____
   causing this fear. When did you
   first become aware of this fear
   of flying?

3. Have you looked at another 3. _____
   means of getting to St. Louis so
   you won't have to fly?

4. That sounds painful. I'd like to 4. _____
   hear about your job situation
   and your relationship with your
   boss.

CLIENT 6, A 65-YEAR-OLD WIDOW: My children don't understand. They question me about everything—where I go, what I do, who I'm dating. They think I've flipped.

Underlying message: _____

Responses   *Type and Level of Response*

1. Tell me more about your rela- 1. _____
   tionship with your children.

2. Have they always been this 2. _____
   concerned about your welfare?

3. Describe one or two things that 3. _____
   seem to irk them.

4. I'd be interested in knowing 4. _____
   why they appear to be so concerned with your actions lately.

CLIENT 7, A 21-YEAR-OLD MAN: I'm so confused. Here I am engaged to a wonderful girl. Everything is going my way, and yet I'm not content with myself. I feel as if there's something missing in my life.

Underlying message: _____

Responses   *Type and Level of Response*

1. You can't allow yourself to feel 1. _____
   good about yourself, and you're
   wondering why. Tell me more
   about yourself.

2. Elaborate on this a little more 2. _____
   so I can try to understand what
   it is that is making you so
   discontented with yourself.

3. When did this feeling first come 3. _____
   to you?

4. Your job, your family, your   4. _____
   recreational activities, your relationship with your girlfriend are
   all going well and yet . . .

CLIENT 8, A 30-YEAR-OLD MOTHER: I can't cope with my children anymore. They come home from school, turn on the radio, tie up the phone, and
totally ignore me when I speak to them. We fight all day long, and my nerves
feel tied up in knots.

Underlying message: _____

Responses   *Type and Level of Response*

1. How long have you felt that   1. _____
   your situation is out of
   control?

2. The only way they pay attention   2. _____
   to you is when you fight . . .
   and you feel anxious about this
   and unable to change things.
   How does your husband
   manage with this?

3. You feel you have lost control. I   3. _____
   would like to hear about the
   times when you have enjoyed
   your family.

4. Please tell me more about your   4. _____
   home life—your daily routine,
   the makeup of your immediate
   family.

CLIENT 9, A MIDDLE-AGED MAN: Mary and I have been married for twenty
years. It was really good in the beginning, but these past few years our marriage
has been terrible. I'm considering a divorce. I just can't take it anymore, it's not
worth it.

Underlying message: _____

Responses   *Type and Level of Response*

1. It must be very painful for you   1. _____
   to live as you do right now.
   What have you done to modify
   or change the situation?

2. You can't take it, but something   2. _____
    tells you that you value or
    cherish something in the rela-
    tionship. Tell me about those
    things that you value in this
    relationship.

3. You seem to keep asking your-   3. _____
    self, "How can something that
    was so sweet become so
    bitter?"

4. Is this feeling only yours? Do   4. _____
    you think Mary feels the same
    way?

CLIENT 10, A 28-YEAR-OLD WOMAN: I'm depressed all the time. Everything I do turns out to be a failure. My family criticizes me. My boss yells at me. Nothing is right for me. Please help me.

Underlying message: _____

Responses   *Type and Level of Response*

1. It's lonely having to get   1. _____
    through your day with no sup-
    port. Is there anyone you can
    talk to?

2. As I listened to your feelings of   2. _____
    failure and loneliness, I was
    wondering about the cir-
    cumstances that surround your
    family's and boss's criticisms.
    Tell me about them.

3. You feel like quitting your job   3. _____
    and moving away from your
    family, but you're wondering if
    that would really help.

4. I'd like to know how you re-   4. _____
    spond when you are criticized
    by your family or yelled at by
    your boss.

COMMUNICATION EXERCISES. Make two different high level probing or inquiring responses for each of the following ten excerpts. Try to use open-ended or probing statements, and be spontaneous in composing your replies.

Client 11, a young woman, 17 years old: It's my father. For years now, I can't remember the last time he came home sober. He's drunk all the time now. He thinks that he's fooling everybody, but he isn't. All my friends know what's going on. I can't take it anymore.

Client 12, a 25-year-old woman: I still can't believe I'm divorced. I don't have anyone to share my life with now. And I have to start the dating scene all over again. I don't know what I'm going to do!

Client 13, a 40-year-old woman: My problem is that I have twin boys age seven, and I can't find the energy or patience to be with them anymore. I wish I were back at work—they're driving me crazy. I feel so guilty about this.

Client 14, a high school senior: I think I'm going to quit school. I can't stand the courses. I don't feel I'm learning anything.

Client 15, a 50-year-old man: I caught my daughter smoking marijuana, and I've put her on restriction. It really hurt me, and now I'm wondering what else she might be doing.

Client 16, a 23-year-old man: I'm thinking that I've made the biggest mistake of my life. I got married two months ago, and now it's not working out. We live upstairs from her parents. My wife and I fight constantly, and her family always takes her side.

Client 17, a 23-year-old homemaker: My drinking is starting to frighten me. It seems like I'm climbing into the bottle more and more often.

Client 18, a 22-year-old college student: In two months I graduate from college. I don't have a job to go to or any plans, for that matter. I don't really want to move back home, but there's nothing else for me to do. I just don't know.

Client 19, a high school freshman: I feel very lonely. I don't seem to have any friends. When lunchtime comes I eat by myself. I don't see anyone after school.

Client 20, a 21-year-old man: I'm twenty-one years old and am basically dissatisfied with my life. I'm not happy with my job—I'd like to go back to school, but I don't know for what. I'm wasting my time, and I see my whole life as meaningless.

SOLO EXERCISES. Develop your probing skills by trying them out in a triad. Each counselor-in-training should practice for at least 10 minutes using all five counseling roles. Each "supervisor" should take the responsibility for criticizing the minicounseling sessions. Repeat the exercise until all members of the triad have had the opportunity to practice.

THE MEANING AND USE OF SILENCE

Beginning counselors often do not know how to deal with silence in the counseling process. For counselors-in-training it can be a frightening experience

(Hackney & Cormier, 1988). In the ordinary social communication process, silence is often interpreted as a negative response. If a person says something, he or she normally expects a response. However, silence is a very effective tool in a counselor's repertoire. Used correctly, it is an active and positive response mode that fits into any number of the counselor's role functions. The use of silence requires a detailed discussion and explanation.

First, silence actively demonstrates your capacity to attend and to listen to a client. Used as an attending response, silence can show that you are sincerely interested in your client and what he or she has to say. This receptive role provides the client with the opportunity and indeed, in some cases, the pressure to speak about and focus on or develop his or her problem. Used in this way, silence conveys the message "I care about you and I am interested in what you have to say." According to Okun (1987), there are times when silence is the only effective way to attend to a reluctant client.

Second, silence can also show support for the client and provide motivation for the client to speak. Silently waiting for another person to speak indicates that you believe that the client is a significant person and worthy to be heard. For clients who are shy or less articulate, silence can actively show openness and respect and provide space for a client to speak. Silence also communicates to clients that they have the responsibility for major inputs in the counseling process. As a counselor-in-training you will need to learn to use silence in this motivating way and avoid the temptation to talk or fill in to remove pressure from the client.

Third, your silence provides an opportunity for your client to clarify his or her thoughts and feelings. This reflective use of silence allows your client to sort out, think about, and reflect on what has occurred so far in the interview. Your client will periodically need to stop, observe what is going on, and gain some insights into his or her progress in the counseling session. This use of silence allows your client the space for his or her own growth-producing thoughts (Evans, Hearns, Uhlemann, & Ivey 1989).

Fourth, silence can be used in the probing or inquiring mode. This inquisitive use of silence can be employed when you actively encourage the client to elaborate on a topic; focus on or delve deeper into a particular thought, feeling, or action; or perhaps weigh alternative courses of action. Silence used in this way communicates that more client information, thought processes, or insights are to be developed and expressed. Using silence as a quest for information reinforces to clients that they have responsibility for progress in the counseling session.

Finally, silence is used in the restive sense. This use of silence occurs when either you, the client, or both of you are intellectually or emotionally fatigued, or when the session has moved too quickly and a pause to rest is needed. You can employ this response mode to slow down the pace of the interview.

Silence is not an indiscriminate tool that you can use in an unsystematic, passive way. It must be used at appropriate times—that is, when it will enhance the role functions mentioned previously. The indiscriminate use of silence often reveals a counselor who is passive and reactive rather than dynamic and active.

This passive silence is the hallmark of a counselor who takes minimal responsibility for the therapeutic process. Excessive use of silence runs the same risk.

Beginning counselors often use silence unintentionally rather than indiscriminately. Its use is rarely purely unintentional, for it does provide the beginning counselor with a reflective time to ingest the client's internal frame of reference and to try to determine what to do next. However, this unintentional use of silence frequently serves concurrently as another role function for the client and hence can be a productive response for the client as well as for the counselor.

As a beginning counselor, you should also be aware of the reasons clients use silence. "Nonverbal client statements" are used when clients are in four different conditions. Silence is used when clients are in resistive, reflective, inquisitive, or exhaustive states.

The client's resistive state may be caused by pain and discomfort or anger and hostility. In the former case, the client may find it hard to discuss something because it causes uneasiness or embarrassment. The client may not feel comfortable enough in the counseling relationship to reveal more about himself or herself (Brammer, 1988). In the latter case, the client may not want to discuss some aspect of the problem—the pain or discomfort may be too difficult to come to grips with in the session. When resistance is encountered, the relationship has not been well established, and as the counselor you should focus on improving the relationship to enhance the trust between you and the client.

The client is in a reflective state when he or she is silently pondering something during the process of counseling. During this time of reflection, the client may be reviewing what has just occurred in counseling, thinking about what he or she wants to say, searching for some information that is not immediately at the conscious level, or solving some internal problem that may lead to some insight or step in the problem-solving process. This reflective use of silence is labeled *integration silence* by Hackney and Cormier (1988) because clients use this time to absorb what is going on in the counseling session. This reflective state is normally quite productive, and you should allow the client ample time to deal with these inner thoughts before trying to proceed with any verbal responses.

The client is in an inquisitive state when he or she is confused and awaiting some action on the counselor's part (Benjamin, 1987). Normally, the client is waiting for some information, support, evaluation, or assistance in the problem-solving process from the counselor. You need to respond to this request for help in a way that will minimize a dependency relationship with the client. Frequently, it is in the best interest of the client and the counseling process to provide the information and assistance that is requested. However, to foster the client's sense of responsibility and to help the client learn how to do certain things, it is sometimes best to help the client learn how to obtain the information and assistance from others.

The client is in a restive, or exhaustive, state when he or she is intellectually or emotionally worn out and needs a moment to catch a breath before responding verbally. When this occurs, you should allow the client sufficient time and wait patiently until the client is ready to proceed.

Because most counselors-in-training have not learned how to use silence effectively in the counseling process, it often leads to rather awkward experiences for beginning counselors. Learning how to use silence requires you to understand what clients are communicating by their pauses and to employ silence as an effective response. To practice being more comfortable with silence, try the following exercises:

1. Form a triad. After the "client" and the "counselor" have started a practice session, the client should try to use silence in one of the four ways outlined in this section. The "supervisor" should monitor this carefully and discuss the client's meaning when appropriate.
2. Form a triad. After the "counselor" and the "client" have started a practice session, the "counselor" should try to use silence in one of the four ways outlined in this section. The supervisor should monitor this session and stop the session when necessary to discuss how silence was used.

SUMMARY

The intermediate role communication skills were presented in this chapter in a manner that required your active involvement in the learning process. After reading and practicing the skills outlined in the chapter, you should have a good understanding of the informing role, the probing role, and the effective use of silence. Furthermore, you should be able to identify these skills, explain the differences between higher and lower level responses, and employ these skills appropriately in your counseling interviews.

The informing or describing role is used when you need to provide the client with relevant facts or descriptions of how various entities function. Informing responses are frequently employed for one of the following reasons: to structure the counseling process, to provide the client with appropriate factual information, to inform the client about some psychological principles, or to present information about a particular intervention strategy. In high level responses the material presented is necessary and appropriate to advance the counseling process and the descriptions are factually accurate; these responses are presented in a manner that is understood by the client.

The inquiring or probing role is employed when you seek to obtain information from the client. Probing responses are typically used to open an interview or to encourage clients to elaborate on a topic; to discuss a thought, feeling, or behavior more completely; to provide concrete examples of their concerns; and to identify their internal or external resources. Inquiring responses include open-ended questions, open-ended statements, and close-ended questions. High level responses focus on critical issues or situations that need further elaboration and discussion, and they demonstrate a sincere and genuine interest in the client. To avoid excessive questions in an interview and to enhance the client's sense of responsibility for the counseling process, it is recommended that the statement form of the probe be used whenever possible.

Silence can be an effective communication skill and can serve a variety of functions in the counseling process. It can be used to show interest and active listening, to show support and concern, to provide clients with the opportunity to clarify their thoughts and feelings, to probe, or to provide time for reflection.

The skills presented in this chapter require extended practice to be mastered. Many counselors-in-training have practiced the statement form of the probe in a variety of situations and have found it to be quite helpful in enhancing their conversations. Because informing, probing, and silence are important communication skills in the counseling process, it is important for you to practice them until you feel comfortable using them.

REFERENCES AND SUGGESTED READINGS

Benjamin, A. (1987). *The helping interview* (4th ed.). Boston: Houghton Mifflin.

Brammer, L. M. (1988). *The helping relationship* (4th ed.). Englewood Cliffs, NJ: Prentice-Hall.

Cormier, W. H., & Cormier, L. S. (1991). *Interviewing strategies for helpers: Fundamental skills and cognitive behavioral interventions* (3rd ed.). Pacific Grove, CA: Brooks/Cole.

Egan, G. (1990). *The skilled helper: A systematic approach to effective helping* (4th ed.). Pacific Grove, CA: Brooks/Cole.

Evans, D. R., Hearn, M. T., Uhlemann, M. R., & Ivey, A. E. (1989). *Essential interviewing: A programmed approach to effective communication* (3rd ed.). Pacific Grove, CA: Brooks/Cole.

Hackney, H., & Cormier, L. S. (1988). *Counseling strategies and interventions* (3rd ed.). Englewood Cliffs, NJ: Prentice-Hall.

Hutchins, D. E., & Cole, C. G. (1986). *Helping relationships and strategies.* Pacific Grove, CA: Brooks/Cole.

Ivey, A. E., Ivey, M. B., & Simek-Downing, L. S. (1987). *Counseling and psychotherapy* (2nd ed.). Englewood Cliffs, NJ: Prentice-Hall.

Johnson, D. W. (1990). *Reaching out: Interpersonal effectiveness and self-actualization* (4th ed.). Englewood Cliffs, NJ: Prentice-Hall.

Okun, B. F. (1987). *Effective helping: Interviewing and counseling techniques* (3rd ed.). Pacific Grove, CA: Brooks/Cole.

Patterson, L. E., & Eisenberg, S. (1983). *The counseling process* (3rd ed.). Boston: Houghton Mifflin.

9

Advanced Role
Communication Skills

INTRODUCTION

This chapter delineates advanced communication skills: the motivating or prescribing role, and the evaluating or analyzing role. The purposes that these skills are used for are given, the types of responses that are used in each of these roles are outlined and illustrated, and practice exercises are provided. You should be able to respond to the following discussion questions after studying this chapter:

- Summarize the purposes of the motivating and evaluating roles.
- Discuss the characteristics that distinguish high level responses from low level responses in each of these roles.
- Identify the types of responses used in the motivating and evaluating roles and give examples of good and poor uses of these responses.
- Apply these advanced communication skills in appropriate situations.

THE MOTIVATING OR PRESCRIBING ROLE

Motivating the client to act is a function of the entire counseling process, and hence this role is implicit in every response or statement you make. Nevertheless, there may be times when you need to employ a role that is more explicitly motivational. This role is manifested when you use direct, deliberate, focused, and forceful statements that are specifically employed to initiate some growth, movement, or productive action on the part of the client. The use of this role requires you to take a different tack in the counseling relationship from the previous roles, and it involves some degree of risk. This type of response is ordinarily used only after a good working relationship has been established. This will allow you time to obtain sufficient information about the client, to determine whether a motivational response is appropriate, and to use this response to strengthen rather than weaken the relationship. The role is normally undertaken to overcome some impasse or to encourage the client to think, feel, or behave in some new or different way. You need to be sensitive to your client when using these responses so you can monitor their effects on your client.

Responses that are used in the motivating role can vary in their potency and strength from relatively mild to extremely strong. If you think that the relationship is not sufficiently well established or that the client is unable to respond in an appropriate way, then you should give some thought to employing milder forms of the motivating and prescribing statements. If the relationship is solid, then you may wish to use stronger forms of these responses.

Types of Motivating or Prescribing Responses

Although a variety of verbal responses can be used in this role, the following ones are the most common:

1. *Advising or directing.* This is a response that recommends a solution to the client's problem or instructs the client to take some action. These responses are more directive than the information responses discussed in Chapter 8. Giving advice is a fairly common response that individuals use in their relationships with others. However, the use of this response in counseling is controversial, and there are several pitfalls that can make its use counterproductive (Brammer, 1988; Evans, Hearn, Uhlemann, & Ivey, 1989; Johnson, 1990). The major difficulties in using an advising or directing response are that it may encourage a client to avoid taking responsibility for his or her actions, enhance the client's feelings of inadequacy, foster a dependency relationship, be trival, or be premature. Furthermore, your client may not follow your advice or recommendation because it may be something the client cannot or will not do. The client may not see the sense of it, or the advice or recommendation may not agree with the client's beliefs or opinions about the topic.

When this role is used, the advice or direction should be given only after very careful thought and when it is meaningful to the client's frame of reference, necessary to advance the client's progress, very likely to be used, and given with directions that are based on solid knowledge (Benjamin, 1987; Cormier & Cormier, 1991). Care must be taken when giving advice to ensure that your client does not forgo his or her responsibility for resolving the problem and that the responsibility for growth does not shift to you. These responses vary in their potency and activity from a *mild suggestion* to "watch something" to a *very directive command* to "do something."

2. *Focusing.* This directive response is designed to concentrate the client's attention on the issues that you as the counselor perceive to be most important at that moment of the interview. This response can be used very effectively when the client presents a confusing or vague situation, rambles on with too many details or too much tangential information, wanders away from an important topic, or provides an incomplete or confusing picture (Brammer, 1988). The response is ineffective when a counselor focuses on irrelevant data or fosters

further digressions. Attending, clarifying, or probing responses can be used as focusing techniques when the purpose is to have the client look at or center on a core notion that you believe to be critical. The responses can vary from a relatively *mild reminder* to keep the conversation on track, to a *strong request* to discuss one's feelings in further detail, to a *command* to elaborate on the meaning of one's nonverbal behaviors.

3. *Confronting.* The purpose of a confrontation is to have a client face up to something that he or she may not be fully aware of or may wish to avoid. It is an extremely useful response that can challenge your client to examine his or her own statements and behaviors more thoroughly (Hackney & Cormier, 1988). Confrontation can be used very effectively when a client reveals discrepancies and contradictions in thoughts, feelings, or behaviors; manifests rationalizations and other poor coping strategies; employs mixed messages; or demonstrates differences between his or her personal and social values (Evans, Hearn, Uhlemann, & Ivey, 1989). A confronting response may also offer a point of view that is different from the client's in order to have the client understand other positions or objective reality (Brammer, 1988; Okun, 1987).

Because a confrontation may deal with matters that are anxiety provoking to the client, they should be given only after a good relationship has been established and be stated in a matter-of-fact tone. Confrontations can be counterproductive when they sound accusatory or judgmental; when they are stated in a blunt, harsh, or critical manner; and when too many are given in one session (Cormier & Cormier, 1991).

A typical confrontation is an honest and direct statement, such as "You said this; however, the evidence suggests something else" or "Your words and your actions do not agree with one another." When the available evidence appears to be contrary to what the client has said, the confrontation may be quite sharp. This "shock treatment" may, as Okun (1987) has indicated, get the client off dead center. However, it may be interpreted as criticism and upset the client. This sharp confrontation has a high degree of risk, and you will need to exercise caution to ensure that it will lead to the desired outcome.

4. *Self-disclosing.* The purpose of self-disclosure is to help the client focus and expand on an issue by offering a germane personal illustration of a similar situation or concern. If you have a personal experience that can help a client, Egan (1990) believes that sharing this information should be a matter of common sense. There are advantages and disadvantages in using self-disclosures in counseling. On the positive side, they have the possibility of improving the relationship, enhancing the client's feeling of trust, helping the client understand that other human beings have similar kinds of problems, motivating the client to share feelings and personal concerns, and helping the client gain another perspective on an issue or learn how to handle a specific situation (Cormier & Cormier, 1991). On the other hand, self-disclosures have the negative possibilities of shifting the focus of the discussion from the client to you, the counselor; being irrelevant to the topic or to the client's perceptual world; or trivializing the client's experience (Evans, Hearn, Uhlemann, & Ivey, 1989).

Self-disclosures are often quite helpful when a client has recently had a rather traumatic experience that you have also experienced. Used in this type of situation, your self-disclosure can be a manifestation of your empathic understanding of the depth and seriousness of the client's problem.

When used appropriately, your response should share relevant personal feelings and experiences, relate these experiences implicitly or explicitly to the client's concerns or point out their differences, and maintain the focus of the interview on the client. Egan (1990) points out that self-disclosures can be a form of modeling and can be used effectively with a client who is reticent and reluctant to talk about self. Your self-sharing will show the client how to talk about self and will encourage the client to reveal more about himself or herself.

In order not to distract from the client's concerns and to keep the focus on the client and not on you, the counselor, self-disclosures should be selective, focused on the issue, and used relatively infrequently and only after a good relationship has been established (Egan, 1990; Evans, Hearn, Uhlemann, & Ivey, 1989).

5. *Making referrals.* One particular type of advisory or directive response is the referral. Brammer (1988) has indicated that referral is an important skill for the experienced as well as the novice counselor. Chapter 3 explained how referrals are designed to help the client in two distinctive ways: first, when the client needs to obtain specific facts or information that he or she will bring back to the ongoing sessions to help the counseling progress; and second, when you feel the client can benefit from another helping professional who has some expertise needed by the client. Making an effective referral requires you to know what resources are available, determine the client's needs and readiness for the referral, select an appropriate and suitable one, and explain to the client in concrete and realistic terms why the referral is being made. It is most important for you to maintain your relationship with the client to ensure that the referral is appropriate and useful.

Different Types of Motivating or Prescribing Responses

The following three examples illustrate the different types of motivating or prescribing responses.

CLIENT 1, A 14-YEAR-OLD GIRL: My father and I fight all the time, especially about the boys I know. He won't even let a boy in the house. He says they're all no good. I think he just doesn't trust me.

1. *Advising or directing:* Instead of attacking your father, you might talk to your mother and ask her if she can help you improve your relationship with your father.

2. *Focusing:* You feel that he doesn't trust you. But it sounds to me that the issue is not just one of trust—you also cannot talk to your father.

3. *Confronting:* You have some rights, but at your age your father also has something to say about your friends.
4. *Self-disclosing:* I remember when I was fourteen—I thought my parents' attitude was positively medieval. I know times are different now.
5. *Making referrals:* Some of the pain you're dealing with might be better handled by an expert in family counseling. I will be happy to talk with you for a while, and we can see if there's a problem we can work on, or if your concern needs a specialist's approach.

CLIENT 2, A MIDDLE-AGED MAN: All my life I've never been able to hold a job for very long. Everything goes okay for a while, but then I get into a fight with the boss or something like that and I'm fired. I can't seem to help it.

1. *Advising or directing:* Instead of allowing yourself to get into arguments with your boss, you should learn some alternative ways to control your temper while at work.
2. *Focusing:* Tell me, what were the circumstances like when you lost your last job.
3. *Confronting:* You know that fights with bosses are unwise, yet you continue to do it.
4. *Self-disclosing:* I've had trouble with bosses also. I don't like to be pushed around either.
5. *Making referrals:* Before we go further in our counseling, I want you to see a psychologist for some tests. These tests will help him, and ultimately you, uncover some of your personality needs.

CLIENT 3, A MIDDLE-AGED WOMAN: My mother died a few months ago, and I haven't been the same since. I just can't seem to get my life back on track again. I've been very preoccupied with death and dying.

1. *Advising or directing:* First we need to help you mourn, to find some time during the day for your crying. Then we can plan on getting your life reorganized again.
2. *Focusing:* Tell me what your day was like yesterday and then how you would have liked to have changed it.
3. *Confronting:* You appear to be using your mother's death as an excuse for ignoring the needs of others in your life.
4. *Self-disclosing:* When my mother died it took me six months to regain my daily routine. It might take about the same length of time for you.
5. *Making referrals:* There are three things that may be helpful. First, I think you should consult your physician about some medication for your nerves. Second, I want you to read *On Death and Dying,* and third, I would like you to join a discussion group I organized. It's made up of people like you who have suffered a loss recently.

Levels of Motivating and Prescribing Responses

Motivating responses, if not carefully used, can put clients on the defensive. This is counterproductive because clients may become indecisive and resistant. Care must be taken not to make clients feel inferior, but rather to make them feel involved and responsible for the outcomes of the counseling process. High level motivation responses are based on the client's need and readiness for a motivational or prescription statement; focus on specific major issues raised directly or indirectly by the client; are usually stated in a warm, friendly manner; offer suggestions for clients to examine their behavior, feelings, or situations in new ways; and encourage clients to become more aware of themselves. Normally this role is used after a good relationship has been established, and it is used sparingly during any interview. The effectiveness and appropriateness of a particular motivational response is a function of its timing and the connotation it conveys. The role is used at all counseling stages; however, it is normally used moderately in the earlier stages and more extensively in the later stages. The use of this role can be ranked on the following four-point scale:

Level 1: Poor use of the role. The counselor is ineffective in using this role when it is used prematurely; when it is blurted out without reference to the needs of the client; when it is presented with an intonation that communicates another role; or when it is conveyed in an accusatory, judgmental, or hostile manner.

Level 2: Mediocre use of the role. The counselor is somewhat effective in using this role when directions are unclear, confrontations are vague, voice intonation is not in tune with the message, and the counselor seems to be unsure of what direction to move the client or the process.

Level 3: Good use of the role. The counselor is effective in employing this role when advice and directions are sound, confrontations are focused or timely, the role is used at a fitting time in the relationship, and the counselor responds properly to the needs of the client.

Level 4: Excellent use of the role. The counselor is most effective in using this role when he or she is aware of the need to use forceful statements to obtain some progress within the relationship, demonstrates sensitivity to the client, and gives an impetus to move the counseling process along.

Different Levels of Motivating and Prescribing Responses

The motivating or prescribing roles are illustrated in the following three examples. The Level-of-Response Scale was used to rate each of the counselor's responses, and the reasoning behind each of these ratings is stated. In studying these examples, you may find that you do not agree with these ratings. Because tonal quality and other paraverbal aspects of the client's statements and the counselor's responses are not conveyed, it is possible that you will read some of these statements in another way than the person who rated them. Any response may obtain a different rating depending on the paraverbal that the reader hears. Thus, it is crucial to indicate the reasons behind the ratings so that any differences among these ratings can be understood.

CLIENT 1, A COLLEGE SENIOR: I'm so mad at the dean. He just asked why I took part in that demonstration. He acted as if I did something wrong.

Responses		*Ratings and Reasons*	
1.	Consider making an appointment with the dean and explain the reasons for your participation to him.	2.3	Although this advice appears appropriate, it may be premature, and it takes the focus off the client's ability to deal with the frustration.
2.	You are *really sure* you were right in acting the way you did.	2.8	The counselor is trying to confront the client in an attempt to have the client focus on her inner feelings.
3.	When I was an undergraduate, I led several student boycotts.	1.0	Poor self-disclosure—not shown to be related to the client's problem; leaves focus on counselor.
4.	Let's look at what happened during your meeting with the dean more carefully.	3.0	Counselor directs the client to focus and describe the situation.

CLIENT 2, A COLLEGE SENIOR: John and I have had a serious relationship for over a year, and he suddenly tells me he's going to live in Chicago to take care of his father's business—he left just like that.

Responses		*Ratings and Reasons*	
1.	You seem to have a choice of either trying to forget John as best you can or talking to John and finding out where he's at in terms of your relationship, then going from there.	2.5	Provides client with good advice, but takes the responsibility off the client's shoulders and runs the risk of becoming a counselor-dominated session.
2.	I was engaged to be married once. She gave me the ring back the week before graduation. Although I realize my	2.5	Self-disclosure has some danger of not being germane, but the counselor relates it back to the feelings of the client.

experience is different from
yours, I sense the frustration
and anger that you feel.

3.	What happened the last few times you were together? Were there any signs of his lack of seriousness?	3.0	Counselor focuses on relationship and requests the client to reflect on the meaning of the relationship.
4.	It may have been very serious on your side but not serious at all for him.	3.5	This confrontation response may be upsetting. The counselor is forcing the client to face up to a possible explanation.

CLIENT 3, A MIDDLE-AGED WOMAN: My husband and I are very upset with our daughter. Every time she comes home it ends up in some sort of conflict. She's twenty-three years old and collecting unemployment, and she doesn't seem to be going anywhere. I just wish she would get married.

Responses *Ratings and Reasons*

1.	You and your husband need to talk to your daughter about your expectations of her and, maybe more important, her expectations of you.	2.3	Gives premature advice that may not be relevant to the situation if more information were known.
2.	Give me some specific information about these conflicts. Describe how they get started.	3.0	The counselor is requesting the client to focus on the conflicts in order to gain important information.
3.	Yes, and I know you would be happy if she moved out of the house, far away—wouldn't it be peaceful.	2.8	This response is designed to confront the client and have her come to grips with how she feels about her daughter.
4.	So, it's all your fault. You've been a rotten parent, and it sounds like you're on a guilt trip.	2.8	The counselor confronts the parent about her possible guilt feeling. This is somewhat risky, but it can be productive.

Practice Motivating and Prescribing Exercises: "How Do I Say 'Get On With It'?"

Similar to the previous sets of exercises, this set also provides the same three types of practice: discrimination exercises, communication exercises, and solo exercises. The first two sets of exercises should be completed in the same way as those in Chapters 7 and 8. When you do the solo exercises, try to practice all previous roles.

DISCRIMINATION EXERCISES. For each client statement given below identify the feeling and the content underlying the client's statement, and then specify the *type* of motivating and prescribing response employed. Finally, indicate the appropriateness of the response by rating the *level* of the response. Anticipate your classmates' questions about the reasons for your answers, and be prepared to discuss your reasons in class. The excerpts may have been read differently by your classmates, and hence their ratings may not agree with yours. It is important to understand the reasons why these differences exist.

CLIENT 1, A YOUNG MAN: I think I want to get a divorce. When I married Susan, I thought we had a lot in common, but I found out otherwise. Her values are so materialistic that I find them repulsive.

Underlying message: _____

Responses	*Type and Level of Response*
1. I would like to talk to you about a marriage and family counselor I know who is very good.	1. _____
2. My wife is also quite materialistic, and it's a problem in our relationship. But I see her really appealing traits too, and our relationship continues to grow in spite of the rocky moments.	2. _____
3. Let's explore your feelings about Susan. When did you first begin to feel your outlook on life might be different from hers.	3. _____
4. And you're perfect. Susan thinks you're her knight in shining armor.	4. _____

CLIENT 2, A 35-YEAR-OLD WOMAN WHO RECENTLY LOST HER HUSBAND: It's only when I'm alone that these feelings and memories keep returning. If I don't do something right away to divert my attention, I just start crying and feeling sorry for myself.

Underlying message: _____

Responses	*Type and Level of Response*
1. I haven't experienced the depth of your loss, but I know when my father died, I cried every night for a month.	1. _____
2. And the eleventh commandment is "Thou shall not cry!"	2. _____
3. Next time you're alone and feel like crying, let yourself go and have a good spell.	3. _____
4. There's a group that meets in the village called Bereaved Spouses. I think you may want to find out more about it. You should plan to attend their next meeting.	4. _____

CLIENT 3, A MALE IN HIS LATE 50s: I just got laid off from my job. I worked for the place for eighteen years, and just like that I'm out on the street. It's impossible for a man my age to begin again.

Underlying message: _____

Responses	*Type and Level of Response*
1. Should I call the undertaker or the old folks' home? You must be out of commission.	1. _____
2. After eighteen years, you don't know how to begin. We have a fine job placement program. I want you to enroll in it.	2. _____
3. Tell me what you mean by the phrase "just like that"?	3. _____
4. My father was laid off after twenty-five years with the same firm. I can sense what you're going through.	4. _____

CLIENT 4, A HIGH SCHOOL SENIOR: I'm not sure where I'm at. I'd like to go to college and get married. I think raising a family can be a full-time commitment. But then again I should think of some career or occupation. Suppose I don't marry or can't have kids. I'm kind of muddling through.

Underlying message: _____

Responses	*Type and Level of Response*
1. You will go to college. You'll work for a few years after college, then you'll get married, have several children, and go back to work when the kids are in school. That's the contemporary female career pattern.	1. _____
2. So, who cares. If you want to go to college, you'll get your application in on time. If you don't, you'll diddle daddle.	2. _____
3. I was in the same boat. I was very confused, but I wasn't able to explore my options with a counselor. I think we can help you do the kind of planning that will help you.	3. _____
4. Well, let's try to get a handle on this. Let's talk about one aspect at a time. First, let's talk about your parents' perceptions about your future.	4. _____

CLIENT 5, A 30-YEAR-OLD MAN: Last week I lost my job. My boss told me to do something I didn't think was right. I wouldn't do it, and he fired me. I don't know what I should do—I have a family to support.

Underlying message: _____

Responses	*Type and Level of Response*
1. I had a boss who was awful, very unethical and unprofessional; we never got along. I know my experience is not the same as yours, but I can imagine what you're going through.	1. _____
2. You're paralyzed. You think you're the only father and husband who ever lost a job.	2. _____

3. You paid a high price to live up
to your values, but you would
have paid a higher price in the
long run if you didn't. Let's
look at what kind of a job you
had or would like to find.

3. _____

4. I can help you resolve some
personal conflicts you have
about this experience. But
ultimately I want you to plan to
see a placement counselor to
help you obtain another job.

4. _____

CLIENT 6, A 20-YEAR-OLD WOMAN: Everything was great when I first met
Tom. We had so much in common, and he seemed like the right kind of guy
to marry. But lately we're fighting a lot. I'm starting to see a different side of
Tom, and I don't like it. I still really love him though. He's a great guy. Maybe
we'll work things out when we're married.

Underlying message: _____

Responses

Type and Level of Response

1. Tell me more about the two
sides of Tom that you see.

1. _____

2. Are you desperate? Do you
want to be in the boxing ring
for twenty years?

2. _____

3. I've had the experience of
beginning a relationship and
thinking I knew the person. I
realize that my experience is
somewhat different from yours.
Tell me more about yours.

3. _____

4. Now is the time to resolve
conflicts. Trying to work them
out during marriage is bad
planning.

4. _____

CLIENT 7, A MIDDLE-AGED MAN: My teenage son and I don't get along
very well. I guess one could refer to it as the generation gap, but I'm only trying to do what's best for him.

Underlying message: _____

Responses	*Type and Level of Response*
1. So, you tell him what to do all the time.	1. _____
2. I had a similar problem with my own children. It took me a long time to learn how to work with teenagers effectively. Let's focus on your concerns.	2. _____
3. You're only trying to do what's best for him. Tell me what you do for him; be as specific as you can.	3. _____
4. If you're on his back all the time that might be the reason for your difficulties in communicating with him.	4. _____

CLIENT 8, A YOUNG MAN, 16 YEARS OLD: There's this girl that I like, but I can't ask her out because my friends tell me I'm too short and ugly, and no girl would go out with me . . . I don't know what to do. I really like this girl.

Underlying message: _____

Responses	*Type and Level of Response*
1. When I was your age I was afraid to ask a girl out; I was afraid I would be turned down. My experience was not the same as yours, but I believe we can help you work this out.	1. _____
2. Let's forget about the girl for the moment. I would like you to tell me about your friends and your relationships with them.	2. _____
3. The other counselor, Doctor Gill, runs a group counseling session once a week for students who are having some personal problems. I believe it's called the Shy Guys' Group. I'd like you to go.	3. _____

4. Yes, I know, you're short and   4. _____
    ugly. She's tall and just won the
    Miss Universe contest.

CLIENT 9, A YOUNG WOMAN: If I could tell him, just once, where to get off, I know I'd lose my job. I don't think it would be worthwhile.

Underlying message: _____

Responses   *Type and Level of Response*

1. What would you say to him?   1. _____
    Tell me.

2. He really gets you angry, yet   2. _____
    you don't want to lose your
    job. You need a way to ven-
    tilate your frustrations. Being
    bottled up is not healthy.

3. The solution to your problem is   3. _____
    in your hands, not your boss's.
    He can handle himself. You are
    having problems dealing with
    him.

4. Why don't you antagonize him,   4. _____
    or quit and get another job?
    You need to learn to stand up
    for what you believe are your
    rights.

CLIENT 10, A COLLEGE SOPHOMORE: I just quit my job because I found it too hard to keep up with my studies and work after school. By the time I got home I was too tired to study. My father thinks I'm copping out on my responsibilities, and he's really mad.

Underlying message: _____

Responses   *Type and Level of Response*

1. It's not easy working your way   1. _____
    through school. I know how
    hard it is—I had to do it myself,
    and it was a struggle. I know
    we can help you work this out.

2. I have two children in college,   2. _____
    and it is a heavy financial

burden. Your father only wants the best for you, as I do for my children. You need to understand your father's point of view.

3. I think you need to learn another way to talk to your father. He seems to want to make sure that your time is being well spent. Let's work on improving your communication skills.

3. _____

4. The majority of students who attend here also work, yet there's something different about your experiences. I wonder what that is.

4. _____

COMMUNICATION EXERCISES.   Ten client statements are provided in this exercise. For each excerpt please try to make three different high level motivational responses.

Client 11, a college freshman: I came because I had a terrible feeling of depression. My boyfriend, who is nineteen, is threatening to break up with me because I won't have sex with him. We've been going together for sixteen months, and everything was great until about two months ago. Now he has this big thing about going to bed. He keeps saying I'd do it if I really loved him. He doesn't understand my viewpoint at all.

Client 12, the wife of an alcoholic husband: My problem is that my husband is an alcoholic. I'm the sole support of my family, and although I loved him once, and suppose I still do, I don't think that staying with him is helping the children.

Client 13, a woman who has had three miscarriages: I've been so terribly depressed lately that I can no longer function effectively. My husband is considerate and understanding, but he says I must resign myself to the fact that I will never have children. I just can't accept that.

Client 14, a 10-year-old girl: Nobody likes me. I don't have any friends. My mother says not to feel bad. I can't help it if I want other kids to like me.

Client 15, a teenager: When I get home from school my mother is always drunk. The house is a mess. She doesn't do the dishes or clean up, and she hardly cooks any supper. She just lies there and drinks and screams at me, and I scream back. I'm so ashamed that I can't ask my friends to come over, and I'm so disgusted sometimes that I think I'll just leave home.

Client 16, a teenager: I know someone who's selling drugs, and I don't know what to do about it.

Client 17, a middle-aged woman: I'm so depressed. I went to the doctor and found out that I'm pregnant. I feel like this is the end of the world.

Client 18, a foreign student: This is my first year in the United States. I had a scholarship this year, and applied for one for next year but didn't get it. So now I am forced to consider a job. You know I only have a student visa. I am not supposed to work here, and since I don't have a degree from the U.S. it will be very difficult to get a job. What would I do if I couldn't get a job? What kind of life do I expect to lead? How can I go back home without completing my studies? I don't even have enough money to go back.

Client 19, a teenager: Both my parents think that I'm still a little kid. I just turned fifteen, and they won't let me go out on a single date with my boyfriend. All of my friends are allowed to date, so I don't see why I can't.

Client 20, a mother of two young children: My husband had a stroke. He'll be in the hospital a long time, and I need money to support us while he is there.

SOLO EXERCISES. Your minisession should now be expanded to a minimum of 20 minutes; continue to practice each previous role communication skill and try to incorporate the new role into your counseling repertoire. Give each person in the triad a chance to practice being the counselor, the client, and the supervisor. Rate your peers on their responses, and review the previous roles when necessary.

THE EVALUATING OR ANALYZING ROLE

The purpose of the evaluating and analyzing role is to help clients become more aware of themselves and the information they have presented about themselves, conceptualize and crystallize the patterns that appear about their concerns, and gain an understanding of the progress they have made during the counseling process.

This role requires you as the counselor to use your professional knowledge of human behavior, your clinical judgment, and your personal sensitivity to assess, analyze, interpret, or integrate various pieces of information that you have obtained about your client. This information may come from signals and messages communicated by the client, psychometric data or other resource material, other significant individuals, or your own professional insights. This evaluation is then shared with your client in order to provide the opportunity for your client to integrate and make associations or connections between and among seemingly unrelated factors, determine meaningful goals, identify resources and strengths, and view himself or herself or his or her concerns in new or different ways. Effective use of this role can facilitate the counseling relationship and can help your client understand possible explanations for his or her behaviors and gain insight or a deeper and better understanding of self.

In using this role, you, as the counselor, assume a more authoritarian, judgmental, and active stance in the relationship. Care must be exercised in using the role because an evaluation process can put your client in a defensive mood. So that your client is not overwhelmed, use interpretations and evaluative responses sparingly. These responses should be reasonably close to the client's previous knowledge or perceptions. Relatively mild forms of interpretations can be expressed by using conditional language, such as, "It seems possible that . . . ," and by emphasizing the client's positive assets, for example, "Your strengths appear to be . . ." Although the evaluative role may be employed at various points in the counseling process, it is most commonly used after the initial stages.

Purposes of Evaluating or Analyzing Role Responses

The evaluating and analyzing role is employed for one of the purposes listed here. It should be noted that these purposes are not mutually exclusive.

1. *Integrating or summarizing a variety of factors.* During the counseling process, a variety of information about the client and his or her concerns can be obtained. This information may have been presented by the client, derived from psychometric testing, given by significant others, or acquired from individuals who made the referral. This response is designed to help the client synthesize this disparate information by providing the client with appropriate feedback, highlighting themes of the problem or the situation, and conveying impressions the client has made.

Effective responses require you to identify the most cogent material, analyze these discrete data, and integrate this information into a coherent picture. This integration can serve as a summarization process to regulate the pace of the interview, give the session a clearer focus, or review the progress of the counseling (Cormier & Cormier, 1991). An effective integration or summarization response should pay careful attention to the major affective and cognitive themes presented by the client (Okun, 1987).

2. *Setting goals.* For the counseling process to be productive, it is important to identify the purpose of this relationship and to establish meaningful goals for the client. Chapter 3 outlines when and how this process should occur. Briefly stated, these goals should be tailored specifically for each client (Brammer, 1988), attainable (Evans et al., 1989), mutually agreed upon (Okun, 1987), and subject to modification (Hackney & Cormier, 1988).

In the evaluative role, you as the counselor should take a directive stance in suggesting what a reasonable goal might be. This step is taken when the client is unable to identify the problem, when you need to translate a vague or diffuse issue into a concrete goal, or when you need to establish one goal when the client has several major concerns.

3. *Identifying the client's resources and strengths.* To help clients understand themselves and take the steps necessary to resolve their problems or concerns, you may need to help them identify their strengths and resources. Identifying your clients' resources includes both internal aspects, such as values, personality characteristics, interests, aptitudes, and skills, and external aspects, such as friends, acquaintances, and the availability of various places and things. Knowledge of these may come from self-analysis, psychometric data, feedback from significant others, and growth or peak experiences that clients have had. In the evaluative role, your task as the counselor is to make your clients aware of these resources as important ingredients in resolving their concerns. You may have to interpret the results of a psychological test, explain how internal strengths can be used to resolve a problem, or explain how a particular external resource can be employed.

4. *Interpreting complex data.* In addition to making evaluative or analytical responses that identify, integrate, or summarize various factors, you may be called upon to explain the meaning of these interrelated factors. The purpose of an interpretation is to teach your clients something about their behavior and to impart some psychological knowledge (Johnson, 1990). This interpretation should help your clients view themselves and their concerns in different ways so they can obtain a better understanding of the problem or situation at hand. This interpretation may describe the meaning of some current event, such as the results of a test, or give some reasons why one behaves in a certain manner (Patterson & Eisenberg, 1983). Interpretations of behavior are frequently given in terms of a particular personality or counseling theory (Brammer, 1988).

Interpretive responses should be used only after a solid relationship has been established and when you have a very good understanding of your client. Counselors who believe that interpretations should emanate from the client and not the counselor do not employ this type of response. Cormier and Cormier (1991) believe that interpretations are generally more effective when they are stated in positive and conditional terms (for example, "Your strengths appear to be . . .").

Different Types of Evaluating and Analyzing Responses

The examples below portray the use of evaluating and analyzing responses. Each of these examples assumes that counseling has progressed for some time.

CLIENT 1, A 25-YEAR-OLD WOMAN: I still can't believe I'm divorced. I don't have anyone to share my life with now. And I have to start the dating scene all over again. I don't know what I'm going to do!

1. *Integrating or summarizing disparate client messages:* As we've discussed your recent divorce these last few sessions, I've been hearing different messages. You've talked about how relieved you are about the divorce and how anxious you are to begin life all over. What I also hear is that the future is

becoming scary for you now that you are alone. I sense that the idea of meeting new people, especially men, is not as exciting as you say it is—your voice even becomes less forceful when you talk about it.

2. *Setting goals:* Before we begin talking about dating, I think it would be important for you to get yourself settled in a job and somewhat over the divorce. That is what you came to see me about. Once this is done, we can explore how you can establish healthy relationships with men.

3. *Identifying the client's resources and strengths:* I can understand your feelings of being lost without having a husband, but I think you're selling yourself short. During the times we've met, you've reflected on how you were the one who really had to make the decisions in your marriage. Also, with your job, you know you can support yourself comfortably. In terms of meeting men, you are active in so many ways that you won't have to get into the bar scene.

4. *Interpreting complex data:* With the points you've just made and other insights you've shared before, I have the feeling you are engaging in a lot of negative self-talk. I sense you are still blaming yourself for the divorce. All the other problems of adjusting, such as living alone and meeting men, are scary because you believe you will fail there as well.

CLIENT 2, A MIDDLE-AGED WOMAN: I have a difficult problem. I went back to work after being home for many years. Now I'm working and making much more money than my husband, and it's tearing him apart. I don't know what to do.

1. *Integrating or summarizing disparate client messages:* From the tone of your voice, I sense you are enjoying your husband's discomfort. At the same time, you want to alleviate the tension your job is causing and restore some harmony to your marriage. To do that you will need to resolve your anger with your husband.

2. *Setting goals:* From the problems you've described, we need to examine your communication patterns with your husband more closely. Once we begin to understand these patterns, we can look at how we might be able to improve your communication skills.

3. *Identifying the client's resources and strengths:* Your concern for your husband is clouding your perception of your abilities to deal with the problem. First, you've shown great compassion and gentleness toward many people. Let your husband receive some so that he knows he's still important to you. Second, you've said your husband has always supported you before, so help him understand why you enjoy this job so that he can begin to support you on this issue as well. Third, you've mentioned how well the two of you can talk with each other and your ability to listen. Listen to him and use this issue to deepen your marriage.

4. *Interpreting complex data:* It sounds like your husband's role as provider has been taken away from him. Your career has progressed very rapidly, while his has reached a plateau. Furthermore, your responses to his behavior appear to reinforce his feelings of inadequacy.

Levels of Evaluating or Analyzing Responses

This role may also be used with different degrees of effectiveness. These responses are most effective when made after the counselor has obtained considerable data about the client and had a chance to integrate this information, and the client has indicated some degree of readiness for an evaluative or analytical response. It is usually desirable to use positive terms to describe the issue, highlight those factors that the client can control, and to employ the role when the client has ample time left in the interview to integrate and respond to an evaluative statement. The counselor who employs the role too frequently in the early stages of counseling may be basing his or her interpretation on unclear connections and fostering a dependency relationship. The role is used inappropriately when it comes into play too early, when it is based on unclear connections, when the counselor's interpretation of the data is not theoretically sound, or when the counselor's body language or tone conveys uncertainty. When used ineffectively, the evaluating or analyzing role puts the client in a defensive posture and discourages the client from revealing more about himself or herself. Its use can be ranked on the following four-point scale:

Level 1: Poor use of the role. The counselor is ineffective in using this role when he or she interprets and evaluates data incorrectly, integrates factual materials based on poor logical connections, stresses negative aspects, underscores uncontrollable factors, and employs the role at the wrong time in the counseling process.

Level 2: Mediocre use of the role. The counselor is somewhat ineffective in using this role when he or she interprets and evaluates only part of the relevant data, does not make appropriate connections among various pieces of disparate data, makes a superficial evaluation, rehashes information, states the negative or uncontrollable factors rather than the positive and controllable ones, or employs the role too early in the counseling process.

Level 3: Good use of the role. The counselor is effective in his or her use of the role when phenomena are integrated or evaluated appropriately, when disparate data are connected logically, when the positive and controllable factors are mentioned, and when it is employed at a suitable time in the counseling process.

Level 4: Excellent use of the role. The counselor is most effective in using this role when he or she interprets or evaluates data skillfully, focuses on relevant issues, makes corrections adroitly, stresses the positive and controllable issues, and employs the role very wisely at a fitting time in the counseling relationship.

Different Levels of Evaluating or Analyzing Responses

To give you practice discerning the differences between levels of responses, three examples are given. Each client statement is followed by a counselor's response. The responses have been rated using the Level-of-Response Scale, and the reasons for assigning these ratings are reported. Examine each of these examples to see if you agree with the ratings and the reasons stated. If you do not agree with these assessments, you may have read one or more of these statements with another intonation than the person who rated them. Because the paraverbal aspects of the client's statements and the counselor's responses are never adequately indicated by the printed word, any response may be rated differently depending on how it is heard. It is very important that you give reasons for your ratings so that you can discuss any differences in class with your supervisor and classmates.

CLIENT 1, A 30-YEAR-OLD MAN:

Ratings and Reasons

COUNSELOR: You seem to have a very strong need to be in control in every situation. When you feel you are losing that control, you become anxious and upset.

3.0–3.5 The counselor summarizes the client's previous statements.

CLIENT: I never really looked at it that way, but now that you mention it, I guess that pretty much sums up the way I am.

COUNSELOR: Wanting to be in control of your life is generally a healthy attitude, perhaps your goal might be to let up a little and not insist on always being in charge of things since this sometimes causes you anxiety.

3.0 The counselor offers some suggestions based on his interpretation of the client's present status.

CLIENT: Easy to say, but how do I actually stop all this control business?

COUNSELOR: The first thing we need to do is try to find out the types of situations that stir up your greatest need for being in control. Is it at work, when you're with your family, or when you're with friends? Then we can get a fix on where your biggest problem lies.

3.0–3.5 The counselor helps the client set some goals to help the client deal with areas where he is experiencing the most difficulty.

CLIENT 2, A COLLEGE SOPHOMORE

Ratings and Reasons

COUNSELOR: Steve, you say you've always disappointed your parents by being more socially oriented

3.0–3.5 The counselor summarizes and interprets the previous statement of the client.

than academically oriented, and that a lot of your resentment is justified because you feel that you were not accepted for being yourself.

CLIENT: You're right. My brother was one hundred percent work, but I'm ninety percent social and ten percent school work. By my senior year in high school, I had become a major disappointment to my parents. This led to their placing restrictions on my social activities, which hurt me.

COUNSELOR: Exactly, and out of this hurt probably came anger, resulting in a knee-jerk rejection of parental directives regardless of their potentially positive effect on you.

3.0 The counselor continues with an interpretation of the client's behavior.

CLIENT: Yes. But I still don't feel that I've been a total failure. My grades are okay, and I think I can get into a decent law school. But dad still thinks of me as an irresponsible kid.

CLIENT 3, A 24-YEAR-OLD WOMAN

Ratings and Reasons

CLIENT: I never seem to be successful in holding a job. I've never had one that lasted more than six months. Everything I do seems to end up in failure.

COUNSELOR: You seem to have had the same problem with each of your former employers. Are you aware that each of your jobs ended in a dispute with your employer?

3.0–3.5 The counselor summarizes and integrates the client's previous statements.

CLIENT: I guess that I never really stopped to think about it, but there were always reasons why I had an argument with the boss.

COUNSELOR: It seems that you have a problem dealing with authority figures in your life, particularly your employer.	3.0–3.5 The counselor becomes more judgmental and interprets the client's behavior.
CLIENT: What can I do about this situation?	
COUNSELOR: We need to explore your feelings about authority figures and set some goals with regard to how you should learn to deal with them in the future.	3.0–3.5 The counselor moves toward setting some goals for the counseling process.

Practice Evaluating and Analyzing Exercises: "How Do I Tell the Client 'These Data Have the Following Meanings'?"

The following set of exercises is provided to help you develop further your evaluating and analyzing communication skills. In each case you should assume that the counselor and the client have had considerable interaction.

DISCRIMINATION EXERCISES. For each excerpt given below, identify the feeling and the content underlying the client's statement. Then state the *purpose* of the evaluating and analyzing response illustrated and indicate the appropriateness of the response by rating the *level* of the response. Be prepared to share the reasons for your ratings with your classmates and your supervisor. You may find that one or more of your classmates read the excerpts differently from you and consequently gave a higher or lower rating to the response.

CLIENT 1, A 19-YEAR-OLD COLLEGE STUDENT: I constantly do things only to gain the approval of others. It seems like I need this praise more than anything else.

Underlying message: _____

Responses	*Purpose and Level of Response*
1. We need to examine your behavior to see why you need that type of reinforcement.	1. _____
2. You have a lot of strengths. During our last meeting you mentioned your excellent grades and your various sporting awards.	2. _____

3. Seeking approval from others is normal. We all need this periodically, particularly from those we love and respect. Your constant quest, however, seems to reveal some unmet needs.

3. _____

4. You mentioned several instances in our last session when you were able to function without this praise. It seems that you only seek it from those who you feel are significant or important to you.

4. _____

CLIENT 2, A 15-YEAR-OLD STUDENT: I just don't understand my parents. Sometimes they want to know everything I'm doing and thinking. And other times they don't seem to care about what I do or what I think.

Underlying message: _____

Responses

Purpose and Level of Response

1. Something has caused this breakdown in your dialogue with your parents. Let's work on trying to improve this communication process.

1. _____

2. Your relationship with your parents is causing you some concern. From what you say, this problem seems to have increased since your mother returned to work.

2. _____

3. Understanding your parents will take considerable time. You do have very good communication skills, and these will be very helpful in improving your relationship.

3. _____

4. You really want to be your own person. You respect your parents, but you don't want to do something just because they said so. Parents of teenagers are hard to understand.

4. _____

CLIENT 3, A HIGH SCHOOL SENIOR: My marks are terrible. I used to be great in school, but lately I don't seem to care. When I try to study my mind wanders, and I can't focus on my work.

Underlying message: _____

Responses	*Purpose and Level of Response*
1. You may have a case of senioritis, but I think your problem has other causes. You've always had the desire to do absolutely great in school. And now you've set standards that you can't meet.	1. _____
2. Let's focus in on that. Tell me what subjects you have difficulty studying and what you do the half hour before you start to study.	2. _____
3. Something recent must have happened to affect your grades. You've been on the honor roll for three years. You proved that you do know how to study and earn good marks.	3. _____
4. And you are concerned that if you keep wandering, you'll be wandering through these halls again next year. The problem seems to have started right after you were turned down by your father's alma mater. That was disappointing for both of you.	4. _____

CLIENT 4, A 30-YEAR-OLD MAN: I can't seem to say "no" to anyone. If someone asks me to do something I'll do it, even though I don't want to. I like myself, but I don't like that part of me that is a gofer for everybody else.

Underlying message: _____

Responses	*Purpose and Level of Response*
1. It's important for you to be liked. That's a learned behavior. You have learned that you tend to service everyone, regardless of the inconvenience that it causes you.	1. _____

2. We need to work on help- 2. _____
    ing you become more
    assertive.

3. You like yourself and you have 3. _____
    a lot of friends. You are capable
    of knowing when you want to
    do something for someone, and
    when you feel someone is tak-
    ing advantage of you.

4. Yet last week you told me 4. _____
    about an incident when you
    didn't go for a cup of coffee for
    your friend. There is some pat-
    tern to when you are a gofer
    and when you are not.

CLIENT 5, AN 18-YEAR-OLD WOMAN: I smoke too much, and so do all my friends. I've tried to stop. I know it's not good for my health, but I do enjoy smoking and so do my friends.

Underlying message: _____

Responses   *Purpose and Level of Response*

1. From the little you have told 1. _____
    me about yourself, you've been
    able to accomplish just about
    anything you've tried.

2. You also told me that you do 2. _____
    not smoke at work or in your
    parents' home. Your habit
    seems to be related to your
    social activities.

3. Let's set a goal for you to stop 3. _____
    in six weeks. I've been able to
    help others within that time
    frame. In order to accomplish
    this, you will have to see me
    each week.

4. Smoking is a learned behavior. 4. _____
    Your smoking has been re-
    inforced by the approval you
    gain from others and the sense
    of relaxation you get from
    inhaling.

CLIENT 6, A 45-YEAR-OLD MAN: I lost my oldest daughter three years ago. She was hit by a drunk driver and was on the critical list in the hospital for five days before she died. It was hard to accept her loss. I know if I hadn't given her permission to go out that night, she would still be with us.

Underlying message: _____

Responses   *Purpose and Level of Response*

1. Grieving is a long and hard pro-   1. _____
    cess that involves several
    phases. You've acknowledged
    the loss of your daughter, but
    you still have to get over being
    angry at yourself. It will take a
    long time for you to move
    through this process.

2. You seem to have some ambiva-   2. _____
    lent feelings about the events
    that occurred before her death.
    While you feel guilty, you also
    said that she was so anxious to
    go out—you could not have
    kept her home.

3. From what you tell me, you   3. _____
    have a lot to be proud of. All
    your children, including your
    oldest daughter, were good
    students and admired by both
    students and teachers. You've
    done a fine job as a parent, and
    you will, I'm sure, continue to
    do so.

4. It often helps to discuss these   4. _____
    important feelings. You have
    kept them inside for three
    years. It sounds like you feel
    you contributed to her
    death.

CLIENT 7, A 55-YEAR-OLD MAN: My wife and I seem to have little in common. For years we were busy raising a family, but now that the children are all young adults we should have time for one another again.

Underlying message: _____

Responses	*Purpose and Level of Response*
1. The problems in your marriage seem to have occurred after the children no longer required your care. It sounds like your family was extremely child centered. Now that focus has disappeared, and your relationship needs to be reestablished on a new plateau.	1. _____
2. I would like to help you improve your communication skills with your wife. You need to learn how to listen to her so that you can understand her needs as well as your own.	2. _____
3. From what you told me in our last meeting, the burden of raising your family was mostly on your wife. You were busy earning money to support them and unfortunately had little time left to help your wife. Now that they are older, you and your wife need some time to adjust to these new roles.	3. _____
4. When you were raising your family, you were willing to put in all the effort the task demanded—you were a fine father and husband.	4. _____

CLIENT 8, AN 18-YEAR-OLD HIGH SCHOOL SENIOR: My boyfriend, John, keeps hinting about getting married. We've been going out, more or less steady, for a couple of years. I don't want to hurt him, but I don't know what I want to do.

Underlying message: _____

Responses	*Purpose and Level of Response*
1. You need to talk more about this. Tell me more about your relationship with John.	1. _____

2. I gather from the little bit you said that you are not ready for marriage. You want to go to college, and you've been accepted at State. It sounds like you want help in learning how to break up with John.

2. _____

3. From what I know about you, you are quite a determined young lady. What you want, you usually go for and get. It sounds to me like you are unsure of what you want to do in this case.

3. _____

4. It sounds like you are unsure of yourself. You are fond of John, and you don't want to hurt him. But you also feel you're not ready for marriage. And you are wondering if you should or shouldn't do what John wants to do.

4. _____

CLIENT 9, A 20-YEAR-OLD COLLEGE SOPHOMORE: I don't know what I really want to do when I finish school. I enjoy my English, history, and philosophy courses, but they don't lead to a specific occupation. I have several friends who are majoring in accounting. Their future is fairly well set, but mine . . .

Underlying message: _____

Responses

Purpose and Level of Response

1. Students who major in one of the liberal arts have a wide variety of career opportunities. Let's begin to explore some of them.

1. _____

2. You mentioned that you were on the dean's list. You not only enjoy your courses, but you also do well in them. You appear to like the academic world. That may be one option for you.

2. _____

3. It's not at all unusual for college sophomores to be unsure about an outlet for their talents. Actually, your vocational development has progressed quite well so far. You've decided to go to college, major in one of the liberal arts, and really master your course work. You do need to do some fine-tuning about further choices.

3. _____

4. You're very wise to think about this now. I'm glad you came to see me. We'll discuss some ways to help you think about your future goals.

4. _____

CLIENT 10, A 16-YEAR-OLD HIGH SCHOOL STUDENT: My father is really upset. He wants me to be on the football team. He thinks it will help me get into a good college. I hate that idea. I never liked that sport.

Underlying message: _____

Responses

Purpose and Level of Response

1. You have many assets that should help you get into college. Your grades are quite good, and you have second chair in the first violin section. Your chances of getting into a good school are excellent.

1. _____

2. You need to work on improving your communication skills with your father. Let's role-play. I'll be you, and you be your dad.

2. _____

3. Last week when you explored colleges your father is interested in, they all appeared to have Division I football teams and outstanding orchestras. Because of your interests, I thought you decided that music was a better option for you than sports.

3. _____

4. Extracurricular activities can and   4. _____
    do enhance the college-
    application process. Your father
    feels you would do well in foot-
    ball, yet you think otherwise.
    Your father's influence is obvi-
    ously causing you some pain.

COMMUNICATION EXERCISES. For each of the following client statements, make two different high level evaluating or analyzing responses.

> *Client 11, a high school junior:* My father spends most of his time on the
> sofa. He usually isn't really bombed—he's just slightly drunk. He's not
> mean or nasty, but he is really an embarrassment. I can't bring any
> friends into the house. I don't know what to do.
>
> *Client 12, a 13-year-old student:* I do just about everything with a group
> of girls in my class. There's one girl who is not part of the group. She
> never was. My friends make fun of her because she doesn't belong. She
> doesn't fit in. They tease her, make fun of her, or ignore her. Sometimes
> she reacts very strongly. I feel sorry for her. I'd like to get to know
> her better, but my friends would get on me if I did.
>
> *Client 13, a college junior:* I think I'm going to fail out of college. I've
> been able to keep up so far, but the work seems to be piling up. I don't
> know what to do.
>
> *Client 14, a 28-year-old man:* I just found out that my wife is pregnant.
> My job doesn't pay very much. I'd like to find another job so I can
> support my family better.
>
> *Client 15, a 20-year-old:* I think a lot about death. My father died last
> year. He was only forty-six. It seems that a lot of people die young and
> leave families that need support. Death seems so unfair and even
> unnatural.
>
> *Client 16, a 45-year-old man:* I'm fed up. I have another new boss. She's
> the fourth one in six years. None of them was any good. I don't know
> how they got their jobs, but my sixteen years of experience on the job
> tells me that many supervisors get promoted beyond their competence.
>
> *Client 17, a 35-year-old woman:* I should never have gotten married. My
> mother said that wives are live-in maids. The children demand this and
> demand that, and my husband never takes a stand. Oh, I wish I knew
> what to do.
>
> *Client 18, a 16-year-old student:* All my friends think I'm strange just
> because I don't go along and do the same things they do. I really don't
> enjoy going to ball games or to the movies that much. I go, and it's
> okay, but I'd rather spend my time doing other things.
>
> *Client 19, a 50-year-old woman:* I don't know what to do with myself
> now that my kids are grown and gone. Nothing interests me anymore.

I don't have the energy or the enthusiasm I used to have. I wish I could find something to do with my time.

Client 20, a 25-year-old college graduate: My wife and I disagree about a family. I would like to have her stop work and have a family. She wants to continue with her job and save some more money. The money's nice, but I know I'll get another raise soon, and with her salary there's not too much left after taxes anyway.

SOLO EXERCISES. Practice all the role communication skills in a triad. Try to conduct a 20 minute minicounseling session. Each counselor-in-training should take turns being the counselor, the client, and the supervisor. Be constructively critical of your classmates' responses.

SUMMARY

The advanced role communication skills presented in this chapter have required your active participation. After reading about the motivating and evaluating responses and completing the structured exercises, you should have developed a good understanding of these skills. You should be able to specify the types of responses used in these roles, discuss the differences among higher and lower level responses, and use these skills in an effective manner in your counseling practice.

The motivating or the prescribing role is employed when the counselor uses rather directive and forceful comments to initiate movement in the counseling process. The following types of responses are typically used in this role: giving advice or directions, focusing on critical issues, confronting, using self-disclosures, and making referrals. High level responses are related to clients' readiness for this type of response; they address the major concerns raised by clients, offer different ways for clients to examine their situations, and encourage clients to become more aware of themselves. This role must be used with caution because it can put clients on the defensive or foster dependency. It is recommended that you use a warm and friendly tone when making responses in this role.

The evaluating or analyzing role is used when the counselor calls upon his or her professional knowledge of human behavior to make clinical judgments and interpret, analyze, and integrate various pieces of information. Responses of this sort typically include integrating or summarizing factors, setting goals, identifying the client's resources and strengths, and interpretating complex data. High level responses require the counselor to have a solid knowledge of human behavior and sufficient information about the client, and a degree of readiness on the part of the client. In this role the counselor must take a very active and judgmental stance in the relationship. To avoid a dependency relationship and to minimize defensive reactions, it is recommended that the responses be expressed in conditional language, such as "It seems possible that . . ." or "Your strengths appear to be . . ."

In addition to reading the chapter and practicing the exercises, you will require considerable clinical practice and critical feedback from your professional peers and supervisors to master these skills. Because these advanced skills are used in many different intervention strategies and approaches, it is important for you to master them as soon as possible.

REFERENCES AND SUGGESTED READINGS

Benjamin, A. (1987). *The helping interview* (4th ed.). Boston: Houghton Mifflin.

Brammer, L. M. (1988). *The helping relationship* (4th ed.). Englewood Cliffs, NJ: Prentice-Hall.

Cormier, W. H., & Cormier, L. S. (1991). *Interviewing strategies for helpers: Fundamental skills and cognitive behavioral interventions* (3rd ed.). Pacific Grove, CA: Brooks/Cole.

Egan, G. (1990). *The skilled helper: A systematic approach to effective helping* (4th ed.). Pacific Grove, CA: Brooks/Cole.

Evans, D. R., Hearn, M. T., Uhlemann, M. R., & Ivey, A. E. (1989). *Essential interviewing: A programmed approach to effective communication* (3rd ed.). Pacific Grove, CA: Brooks/Cole.

Hackney, H., & Cormier, L. S. (1988). *Counseling strategies and interventions* (3rd ed.). Englewood Cliffs, NJ: Prentice-Hall.

Hutchins, D. E., & Cole, C. G. (1986). *Helping relationships and strategies*. Pacific Grove, CA: Brooks/Cole.

Ivey, A. E., Ivey, M. B., & Simek-Downing, L. S. (1987). *Counseling and psychotherapy* (2nd ed.). Englewood Cliffs, NJ: Prentice-Hall.

Johnson, D. W. (1990). *Reaching out: Interpersonal effectiveness and self-actualization* (4th ed.). Englewood Cliffs, NJ: Prentice-Hall.

Okun, B. F. (1987). *Effective helping: Interviewing and counseling techniques* (3rd ed.). Pacific Grove, CA: Brooks/Cole.

Patterson, L. D., & Eisenberg, S. (1983). *The counseling process* (3rd ed.). Boston: Houghton Mifflin.

10

Problem-Solving Skills

INTRODUCTION

The purpose of this chapter is to provide you with the opportunity to practice using the cognitively focused, the affectively focused, and the behaviorally focused intervention strategies that were outlined and explained in Part 2. After studying this chapter you should be able to:

- Describe the cognitively focused, the affectively focused, and the behaviorally focused intervention strategies that are reviewed in this chapter.
- Outline the procedures that you would employ in each of these intervention strategies.
- Indicate effective and ineffective responses in using these interventions.
- Identify some typical situations that you believe are appropriate for the different kinds of intervention strategies.

THE PROBLEM-SOLVING ROLE

The problem-solving role is action oriented. In this role you will use specific interventions to assist your client in learning how to function more effectively. The entire counseling process is, of course, aimed at resolving your client's concern and improving your client's well-being. The various counseling roles discussed previously all facilitate this client growth and assist the client in this endeavor. The problem-solving interventions are used extensively in the working or treatment stage of counseling and include a variety of cognitively oriented, affectively oriented, and behaviorally oriented techniques.

A brief review of these techniques is presented in this chapter. Several examples are then presented to illustrate how a counselor's use of these interventions may vary in effectiveness. Additional examples are then provided to give you some practice in evaluating these interventions. Finally, several exercises are offered to enable you to practice using these strategies.

Types of Problem-Solving Responses

Chapters 4, 5, and 6 presented a more detailed description of each of the major interventions discussed in this chapter. Although a brief review of each of these strategies is presented in the following paragraphs, you may wish to reread these earlier chapters to refresh your memory and recall the methodologies involved in each of these interventions.

Cognitively Focused Interventions

Cognitively focused techniques are based on the premise that individuals are rational beings and hence solutions to their concerns need to focus on the cognitive domain. Clients may need help to attain and retain factual information, make appropriate decisions, or use their deductive and inductive thinking processes more effectively. The cognitively focused interventions are outlined more fully in Chapter 4 and are briefly reviewed here.

HELP CLIENTS OBTAIN RELEVANT FACTUAL INFORMATION. Counselors are often called upon to help their clients obtain important information. This information may be necessary to assist clients in learning more about educational and vocational opportunities, increasing their knowledge of themselves by understanding the results of psychometric instruments, or developing the skills necessary to manage the learning process more effectively.

HELP CLIENTS DEVELOP THE DECISION-MAKING AND PROBLEM-SOLVING SKILLS. Counselors may be called upon to help clients make decisions or resolve problems. Clients may need your assistance in learning how to weigh different facts, resolve internal conflicts, deal with their own frustrations, choose among alternative courses of action, cope with various life problems, or overcome their indecisiveness.

HELP CLIENTS IMPROVE THEIR DEDUCTIVE AND INDUCTIVE THINKING PROCESSES. As a counselor, you may have to help some of your clients understand and overcome an unproductive and debilitating thought pattern and encourage them to learn to think more logically and develop a more constructive and rational thinking process. Other clients may need your assistance in learning to think analogously, inductively, and creatively.

Affectively Focused Interventions

Client concerns are often related to the affective domain; therefore, you may need to encourage your clients to focus on the attitudes, beliefs, and values they have about themselves and others. It is often important to help clients obtain a keener awareness of self, improve their ability to deal with their own emotions, develop improved feelings of self-worth, and gain better acceptance and appreciation of others. The affectively oriented approaches ought to be considered when any of these goals need to be met. These strategies are outlined in Chapter 5 and are summarized in the following paragraphs.

HELPING CLIENTS VENTILATE. Clients may need an outlet to discuss, examine, and investigate their feelings, thoughts, opinions, and experiences in an open, accepting atmosphere. Fostering a warm, accepting climate where clients have the freedom to discuss whatever is bothering them is an important affectively focused strategy.

ASSISTING CLIENTS WITH THE CATHARSIS PROCESS. You may occasionally have a client who has been unable to express a deeply felt emotion. A client who desires to be freed from this burden requires a more complex intervention than pure ventilation. In this situation you will need to help the client release any pent-up tension, bring this emotional blockage out into the open, and thus help the client purge himself or herself of this restrictive feeling.

PROVIDING CLIENTS WITH THE OPPORTUNITY TO IMPROVE THEIR SELF-AWARENESS, INSIGHT, AND SELF-ESTEEM. Still other clients may need assistance in becoming more aware of themselves and the factors that influence their behaviors. You may also want to consider using an affectively oriented strategy when your primary emphasis is helping your clients understand themselves, their relationships to significant others, and the meaningfulness of their own life experiences. Helping clients develop self-esteem is also an important counseling goal. Normally, this requires clients to move beyond the awareness and insight levels in order to reorganize their own self-structures. Accomplishing this goal will necessitate skillful use of an affectively oriented strategy.

Performance Focused Interventions

Clients may present concerns that are related to their behaviors. They may want to learn a new behavior, increase the likelihood of an existing behavior, or eliminate or decrease the likelihood of a present behavior. When this is the case, you should consider using one or more of the performance oriented strategies. These interventions are based on the principles of observational and simulated learning, contingency management, and classical conditioning. The major performance focused interventions are outlined in Chapter 6 and are presented briefly in the following paragraphs.

OBSERVATIONAL AND SIMULATED LEARNING. As a counselor you may have a client who wants to learn a new behavior. When this situation occurs, you should consider employing the two major aspects of observational and simulated learning. First, using the observational learning approach, select a model who can portray the desired behavior and have the client carefully observe the desired behavior in the model. Second, using the simulated learning approach, have the client practice the desired behavior in a simulated manner. The specific techniques that you employ may include role-playing, role reversal, and dialoguing.

CONTINGENCY MANAGEMENT. Counselors are often asked to help their clients increase the likelihood of a specific behavior or decrease the likelihood of an unwanted behavior. When this is the case, you should consider using the techniques based on the principles of contingency management. This will require you to instruct your client on these principles and mutually develop a carefully thought-out plan to implement the steps involved in this approach. You will need to help the client select appropriate positive or negative reinforcers, or choose procedures that may serve as positive or negative punishments.

CLASSICAL CONDITIONING METHODS. You may have other clients who need assistance in dealing with a blockage that prevents them from performing some behavior. When this situation is present, you should consider using one of the intervention strategies based on the principles of classical conditioning. These techniques include stimulus-control methods, relaxation and desensitization methods, assertiveness training, aversion techniques, flooding, and paradoxical intention.

Levels of Skill in Using Intervention Strategies

As with other counselor communication roles, you as a counselor can use this role with varying degrees of effectiveness. On the four-point scale, the use of the role can be rated according to the following criteria:

Level 1: Poor use of the role. A counselor is least effective when he or she attempts to assist the client in an incompetent way or at an inopportune time, and when there appears to be little relationship between the client's concern and the counselor's strategies.

Level 2: Mediocre use of the role. A counselor is somewhat ineffective when an appropriate intervention is made prematurely, when strategies are used only partially correctly, or when strategies are only somewhat useful to the client.

Level 3: Good use of the role. A counselor is effective in using this role when he or she employs theoretically sound techniques, uses them appropriately at a suitable time in the counseling process, and communicates in clear terms.

Level 4: Excellent use of the role. A counselor is most effective when he or she uses the technique that is most appropriate and demonstrates mastery of the strategy by employing it in a timely fashion in the counseling process.

Different Levels of Problem-Solving Responses

The following three cases illustrate the use of the problem-solving role. The counselor's responses have been rated, and the reasons for these ratings are explained. Study each of these examples, examine the counselor's responses, and see if you agree with the ratings and the reasons stated. You may have read the responses with a different tonal quality than the person who rated them and so disagree with the assigned ratings. If you do disagree, please indicate the reasons why you disagree. Be prepared to discuss the reasons for your ratings with your supervisor and your classmates.

CASE 1: Sandy is 26 years old and single. For the past 4 years she has been employed as an office manager in a small firm owned by her family. She has a degree in speech pathology, but she never used her professional training. She came to see a counselor because she wants to change her job and find something she really likes. The counselor has established a relationship with Sandy and has given her the Strong Interest Inventory.

Excerpts and Responses

Ratings and Reasons

COUNSELOR: Sandy, you've had a chance to look over the results of the Strong Interest Inventory. I see that your interests tend to cluster in the social area.

2.7 This appears to be a good response; however, the interpretation of the Strong Interest Inventory is too limited.

CLIENT: Yes, I saw that. When I was studying speech pathology, I really enjoyed my classmates. We had a lot in common.

COUNSELOR: You must have had a good reason not to pursue that as a career, since you put so much time and effort into it.

3.5 This is a good use of the indirect form of the probe. Identifying the reason why a client did not pursue an occupation compatible with her interest is important.

CLIENT: Yes, I didn't for a couple of reasons. First, even though I did quite well, I found that I really didn't like the work involved. And second, there weren't many jobs available at that time.

COUNSELOR: So then you got a job in your family's business and entered the business management field.

2.8 This appears to be a reflection of previously stated information. It may encourage the client to elaborate on her interests a little more.

CLIENT: Yes, but I don't want to stay in this line of work. Office work is not for me.

COUNSELOR: Do you see any relationship between your interests as measured by the Strong and any of your vocational experiences?

3.2 This probe can help the client see if her measured interests are supported by her personal experiences.

CLIENT: Yes, the Strong suggests that my interests are similar to people in the helping fields, and dissimilar to those people in the enterprising areas. I guess that's a reflection of why I liked the people in speech pathology and dislike the work I'm doing now.

COUNSELOR: Your actual or manifest interests were reflected by the scores on the inventory.

3.0 The counselor reflects the content of the previous statement to reinforce this concept.

CLIENT: Yes, but while I enjoyed working with people in speech pathology, I didn't enjoy the actual work that much.

COUNSELOR: In your case, it sounds like we need to help you find an occupation that can meet your interest in the helping process and at the same time has tasks that you find more enjoyable.

3.5 The counselor outlines in very general terms the next steps involved in this career counseling situation.

CLIENT: That sounds great. How do we do this?

COUNSELOR: Are you familiar with the *Occupational Outlook Handbook*?

1.0 Introducing this reference at this point is irrelevant.

CLIENT: No.

COUNSELOR: Well, at some point we may want to look at that. It's a government reference book that describes specific occupations and tries to forecast the need for people in these areas.

2.5 The counselor realizes the inappropriateness of the previous response and explains the general contents of the text for clarification.

CLIENT: I see . . .

COUNSELOR: Well, what I think we should do now is generate a list of potential occupations. Then we can cull from that list the occupations that may be more realistic for you.	3.0	The counselor moves the process along and encourages the client to think about possible choices she has in mind.
CLIENT: Well, I've often thought that I might want to become a special education teacher or a counselor or a psychologist.		
COUNSELOR: Those are three possible choices. In addition to those, we should use the Holland codes on the Strong to help us generate occupations that you might not have thought of.	2.4	Although this is often a good way to generate additional choices, the counselor never fully interpreted the Strong nor explained the meaning of the Holland codes.
CLIENT: I'm beginning to get a little excited about this. It feels good to finally begin to move to do something that I really would like.		

CASE 2: Keith is a 21-year-old senior in college. He came to see the counselor to do some postcollege planning. His grades for the first three years were very good; however, this semester he has not kept up with his studies and his grades have been poor.

COUNSELOR: Hi, Keith, how are you today?	3.0	This is a good opening.
CLIENT: Not too good. As I told you last time we met, I need to improve my grades.		
COUNSELOR: You need to improve your grades.	3.0	The counselor makes good use of the accent to attend to the client.
CLIENT: Yes, I was a solid "A" student and all of a sudden it seems my grades are going downhill.		
COUNSELOR: And you're not sure why this happened.	3.0	The counselor uses the indirect form of the probe to gain further information.

CLIENT: Well, I met a group that likes to party. They're fun to be with, but I'm not spending much time on the books.

COUNSELOR: It sounds like you want to do both—keep going to the parties, but at the same time keep up with your work.

3.5 The counselor uses a paraphrase to communicate back to the client the message that the counselor heard.

CLIENT: Yes, it's hard to do. I tried to stop the partying, but I just can't. I'm really mad at myself.

COUNSELOR: You tried to stop partying, but you can't. There is something there that's too attractive to give up.

3.5 Another good use of the paraphrase.

CLIENT: I don't know. . . . I guess there is.

COUNSELOR: You feel like you're being pulled in two directions. You want to stop partying, yet something draws you to it. You want to resume studying, but you don't have any energy left.

3.0 The counselor tries to clarify between the two alternatives to see if one has a stronger attraction.

CLIENT: I guess I'm trying to play to two different audiences—one wants one play, and the other wants a different one. Which play should I pick?

COUNSELOR: It sounds like you're trying to meet two different needs that you have—one to interact with your peers, and the other to master your work. And you want to satisfy both.

3.5 The counselor interprets the client's problem in terms of his needs and implies that both of these activities have meaning for the client.

CLIENT: Yes, I do. But when I'm trying to study, all it takes is one call from a friend and I run to the party.

COUNSELOR: I'm not sure I understand. Is it the call that makes you go, or is it a fear that you'll miss out on something?

4.0 The counselor uses the clarification response to see if the client takes some ownership of the problem.

CLIENT: I guess it's me. I guess I can't blame the phone call or my friends.

COUNSELOR: Well, there's nothing wrong with spending time with your friends, but it sounds like there's more to this issue.

2.5 This attentive response is fair; it goes over previously mentioned material.

CLIENT: Oh, why is it that I can't seem to say no? Why do I give in to myself?

COUNSELOR: And if we could find the answers to those questions, it would really help.

3.5 Very good paraphrase of the clients statement.

CASE 3: Arlene is a 23-year-old client who lives and works in the suburbs. She has friends who live downtown in a tall building, but she is very reluctant to visit them because she has a fear of elevators.

CLIENT: I can't stand it. Every time I go downtown to meet my friends, I have to ride the elevator to the seventeenth floor. I break into a cold sweat, and by the time I get there I feel faint and clammy all over.

COUNSELOR: This fear really causes you a lot of discomfort.

3.0 This is a good paraphrase of the client's statement.

CLIENT: Yes, it sure does.

COUNSELOR: Perhaps I can help. I would like to know when you first experienced this fear.

4.0 The counselor provides support by offering to help and then probes for important information.

CLIENT: It started about two years ago. I was downtown and alone in an elevator car when it suddenly stopped. It was five minutes before it started up again, but it felt like five hours.

COUNSELOR: You didn't have this fear before this incident.

3.0 The counselor continues to probe for additional information.

CLIENT: No, I was fine, except in one of those high-speed cars where they zoom you to the top floor in seconds.

COUNSELOR: In the last two years you have experienced a lot of pain and anxiety when you ride an elevator.

2.8 Counselor summarizes previous statements. Response may be a bit redundant.

CLIENT: Yes, fortunately I don't have to take an elevator to work. I walk up two flights of stairs— it's good exercise.

COUNSELOR: You could take an elevator at work, but you would rather avoid it and the pain that goes with it.

3.5 Counselor gives a good paraphrase of the client's message.

CLIENT: Well, I do feel some anxiety when I see my friends getting on, but I try to think of other things right away.

COUNSELOR: So even seeing other people walk into an elevator causes you some discomfort.

4.0 The counselor very wisely checks out the importance of this visual stimulus.

CLIENT: Yes, I guess it does.

COUNSELOR: How do you feel when we talk about elevators?

3.5 The counselor continues to probe to see if the anxiety is present when the stimulus is covert.

CLIENT: Right now I feel a bit uncomfortable, but I know I have to resolve this problem.

COUNSELOR: You are determined to overcome this fear.

4.0 Good paraphrase that also supports the client's resolve.

CLIENT: Yes, I really need to.

COUNSELOR: In addition to your daily encounter with the elevator at work, tell me about how often you may need or want to ride an elevator in a typical week.

3.5 The counselor continues to probe to obtain a more complete understanding of the problem.

CLIENT: I would say about two or three times. It's hard to do a lot of things. Can you tell me how long it will take me to overcome this fear?

COUNSELOR: It will take us some
time. There is a systematic way to
help you overcome this fear.
First, I'll teach you a method to
relax. Then we will slowly begin
to talk about elevators. And eventually we should be able to have
you visit your friends on the
seventeenth floor as often as you
like.

3.5   The counselor provides
important information and
briefly outlines the counseling
process for the client.

PROBLEM-SOLVING PRACTICE EXERCISES

"How Do I Say 'I Know I Can Help You; This Is What You Need To Do Next'?"

DISCRIMINATION EXERCISES.   The following examples are excerpts from
several different counseling situations. These exercises illustrate the content,
direction, and type of communication skills that a counselor might use in this
role. They are designed to help you develop your skills in applying a variety
of intervention strategies. For each of these examples, first indicate which intervention strategy is being used; then indicate the appropriateness and the effectiveness of each response by rating the level of the response. Please be
prepared to share the reasons for your ratings with your classmates.

CASE 1:   John is a high school junior who came to see his counselor for
assistance in career planning. He is in a college preparatory program and has
a "B+" average. His test scores indicate that he has above-average ability. His
father is an electrician for a large company and his mother is a homemaker. He
is the oldest of three children.

Excerpts and Responses

Ratings and Reasons

CLIENT: I don't know what I want
to do with my life. I mean, I
don't know what field or occupation I should go in for.

COUNSELOR: Picking out an
occupation seems to be a pretty
scary thing, but you don't need
to make a decision now. What
you should do is begin the exploration process so you can
learn more about yourself and
find out what kinds of
occupations might meet your
interests.

CLIENT: You mean I don't have to decide today or even for a while? How can I start this exploratory process?

COUNSELOR: First, we need to help you get to know yourself better. Then we should take a look at various occupations that seem likely for a person like you. By this process we should be able to eliminate many occupations that are of no interest to you and identify several that you might like to consider.

_____

_____

_____

_____

_____

CLIENT: It sounds like a time-consuming process.

COUNSELOR: Well, it can take some time. But this investment is usually very worthwhile.

_____

_____

CLIENT: Can you describe this process in more detail.

COUNSELOR: Sure, lets break it down into a series of steps.

_____

_____

CLIENT: Yes, I'd like to know the whole process.

COUNSELOR: Well, the first phase is helping you get to know yourself better. To do this, we'll need to take a close look at the things that interest you, your values, and the kind of person you are.

_____

_____

_____

CLIENT: That's fine. How do we do this? Do you want me to take some of those psychological tests?

COUNSELOR: That's one way to do it. We could also use one of the computer programs to help us do some of this.

_____

_____

CLIENT: I'll use the computer.

COUNSELOR: Great. Let's schedule you right now for time to use the program.

    CASE 2: Thomas is 55 years old, and his first wife died four years ago. He has a son 30 years old and a daughter 26 years old. He is currently dating Connie, a widow, who is 45 years old and who has three teenaged children. He is thinking about becoming engaged, but he is very anxious about this move.

Excerpts and Responses   *Ratings and Reasons*

CLIENT: I don't know what I want to do.

COUNSELOR: You sound confused, and you become anxious because you feel you ought to do something.

CLIENT: The situation is complex. I have two adult children, and Connie has three teenagers.

COUNSELOR: In addition to you and Connie there are at least five others who will be influenced by what you do.

CLIENT: Yes, and some others too. My wife's parents and mine are deceased. But Connie's parents have been a tremendous support to her, both financially and emotionally, since she became a widow.

COUNSELOR: She's real close to her folks and has depended a great deal on them.

CLIENT: You have a good picture. The older and the younger generations will influence us in one way or another.

COUNSELOR: I see. Your relationship with Connie will be strongly influenced by both sets of children and by her parents.

CLIENT: I feel like I'm being pulled in two directions. One way I take on a lot more responsibilities, and the other way I can be relatively carefree.

COUNSELOR: Let's try to think this out in a logical manner.

_____

_____

CLIENT: What do you have in mind?

COUNSELOR: Let's write things down—the pros and cons of getting married again. Often when we write things down they become clearer.

_____

_____

CLIENT: I should make a list of the reasons why I should and the reasons why I should not marry Connie.

COUNSELOR: Kind of brainstorm. Do all the cons first; then wait a while and list all the pros. Put the lists away for a while, then take them out and look them over.

_____

_____

_____

CLIENT: I've got to do something. This sounds like a good start.

COUNSELOR: When you've finished, bring in your lists and we'll look them over together.

_____

_____

CASE 3: Karen is a 20-year-old college sophomore who transferred from another college at the end of her freshman year. She had excellent grades in high school and during her first year of college.

Excerpts and Responses   _Ratings and Reasons_

CLIENT: Oh, I can't keep up with the work in Professor Smith's class.

COUNSELOR: You sound angry and upset.

_____

_____

CLIENT: Yes, he has a reputation as a demanding teacher. They say he rejects papers that are not up to his standards.

COUNSELOR: I'm not sure whether it is the content of the course, Professor Smith, or your tendency to be perfect that is bothering you the most.

_____
_____
_____

CLIENT: His standards are so high, and he even wants us to present our papers orally in class.

COUNSELOR: So, in addition to writing a perfect paper, you will have to give a perfect speech.

_____
_____

CLIENT: I can't get his paper finished—it's not right. And he'll be so critical.

COUNSELOR: What is the worst thing that can happen to you if you don't meet his standards?

_____
_____

CLIENT: He'll ask me to rewrite it. But I'll feel awful.

COUNSELOR: You'll feel awful because you didn't get an "F"?

_____
_____

CLIENT: Well, that would be far worse. But he'll still make me give an oral report.

COUNSELOR: What will happen if you stand in front of the class and can't open your mouth?

_____
_____

CLIENT: I'd be so embarrassed.

COUNSELOR: So get a little embarrassed. It's better than a failing grade.

_____
_____

CLIENT: I can't do it—my friends will laugh at me.

COUNSELOR: It sounds to me like "I won't do it because my friends will see that I am a complete idiot."

_____

_____

CASE 4:   Tony is 23 years old and has a degree in art education. He has not been able to find a permanent job in his field and is concerned about his future. He lives with his parents, who would like to see him get established in an occupation.

Excerpts and Responses   *Ratings and Reasons*

CLIENT: My father is on my case. He wants to know when I'm going to use my art degree in a useful way.

COUNSELOR: You sound angry because your father doesn't understand the problems you face in your desire to become an art teacher.

_____

_____

CLIENT: Yes, I can get some work subbing for two or three days a week, but that only pays for my transportation and basic expenses. I can't live on that kind of money.

COUNSELOR: And to your father, that seems like a dead end.

_____

_____

CLIENT: Yes, he feels that after five years of college, I should be able to settle down into a job and begin to earn a decent salary.

COUNSELOR: It hurts for several reasons. First, you have your heart set on working with adolescents and helping them appreciate art. Second, you can't seem to find a permanent position in this field. And then there's your father, who is forcing you to question your commitment to this field.

_____

_____

_____

CLIENT: I know that a job will eventually open up for me. It's hard enough without his pressure.

COUNSELOR: In spite of his pressure and this hurt feeling, you are committed to staying in this field, even if it takes you a long time to find the job you want.

_____

_____

CLIENT: Yes. What can I tell him to get him off my back?

COUNSELOR: It sounds like you're searching for ways to have him ease up on this pressure.

_____

_____

CLIENT: Yes, I sure am. You know he would like me to take a job in his firm. He always wanted me to follow his lead.

COUNSELOR: So part of the problem is related to your choosing an occupation that is different from his.

_____

_____

CLIENT: He's got the ammunition to really give it to me on that issue.

COUNSELOR: And you need to be able to talk to him about what's right for you.

_____

_____

CASE 5:   Lois came to talk to a counselor on the advice of her physician. She is 45 years old and reportedly in good health. She complains that she feels listless and cannot organize her time the way she used to. Her mother died a year ago, and she lives with her husband and two children. She has a good job and is active in several community groups.

Excerpts and Responses   _Ratings and Reasons_

CLIENT: I'm so angry at that doctor. He could have helped Mom more than he did.

COUNSELOR: You really miss your mother, and you are angry with the physician.

_____

_____

CLIENT: He said he could help, but he didn't.

COUNSELOR: It's okay to be angry, but I want you to take some responsibility for feeling this way.

_____

_____

CLIENT: What do you mean?

COUNSELOR: Instead of saying, "That doctor made me so mad," I want you to say, "I am really angry at that doctor."

_____

_____

CLIENT: Well, he makes me mad, and I am really angry with him. I'm sure he could have done more.

COUNSELOR: That's good. You really miss your mother and you blame the doctor.

_____

_____

CLIENT: He said he could help, and I didn't have time to take care of her.

COUNSELOR: So you feel guilty.

_____

_____

CLIENT: Yes, she was such a great mother. She was kind. She was super with my kids.

COUNSELOR: So you feel guilty because your mother raised you, took care of you when you were sick, and helped a lot with your kids. And you were not able to take care of her, so you really feel guilty.

_____

_____

CLIENT: Yes, she was really wonderful. And that cancer—it's so vile.

COUNSELOR: Suppose you could talk to your mom now. What would you say?

_____

_____

CLIENT: How can I do that?

COUNSELOR: Let's pretend your
mother is sitting in this chair.
What would you say to her?

_____

_____

CLIENT: Mom, I miss you. I can't
do what you did. You managed
to do everything so easily, and
I'm always struggling.

COUNSELOR: That's good.
Continue talking to her.

_____

_____

CASE 6: Jim is a high school sophomore who moved into the community six weeks ago. He appears to be doing well academically and has recently
gone out for one of the school's teams. However, he seems to be somewhat shy
and introverted. He would like to ask a classmate out on a date, but he has not
been able to do so.

Excerpts and Responses

Ratings and Reasons

CLIENT: I would really like to ask
Susan to the prom.

COUNSELOR: Well, perhaps I can
help you. Tell me more about
your situation.

_____

_____

CLIENT: I never do anything right.
Every time I see her I get
tongue-tied.

COUNSELOR: You feel as if she will
ignore you or turn you down.

_____

_____

CLIENT: She doesn't ignore me.
She usually smiles and always has
a friendly hello.

COUNSELOR: So, she is friendly
and seems to like you, but you
can't say more than a few words
to her.

_____

_____

CLIENT: I can say hi and talk about
school or things like that, but I
can't seem to get up the nerve to
ask her out. I don't know why.

COUNSELOR: Well, let's role-play that situation. You be yourself, and I'll be Susan. Pretend we're in school and we meet going into a classroom about two minutes early, and I say "Hi."

_____
_____
_____

CLIENT: "Hi, Sue. He sure gave a lot of homework. Could you finish yours?"

COUNSELOR: "Yes, but it took me hours, and I had to get my dad to help. Maybe I should have called you."

_____
_____

CLIENT: If she ever made a statement like that, I'd really get embarrassed.

COUNSELOR: Well, then suppose she did; or is it out of character for her to say that?

_____
_____

CLIENT: No—she hasn't said anything like that, but she could.

COUNSELOR: So, Susan is friendly toward you. But you are afraid of rejection.

_____
_____

CLIENT: I guess so.

COUNSELOR: Let's role-play three different situations. In the first, she says no. In the second, she says she would really like to, but she is busy that weekend. And in the third, she says yes.

_____
_____

CASE 7: Bruce is a 20-year-old college junior who is majoring in mathematics. His cumulative average is 2.95, but he received "D's" on the first quizzes in each of his math classes this semester.

Excerpts and Responses   *Ratings and Reasons*

CLIENT: I need to improve my study habits. With this new major, they really have piled on the work.

COUNSELOR: And you can't seem to get organized.

_____
_____

CLIENT: Yes, I've broken all my resolutions. I find myself reading the *Times* or a magazine before I settle down to study.

COUNSELOR: It's good to keep up with current events and do some leisure reading, but you have to set up some priorities.

_____
_____

CLIENT: Yes, I know—bad marks, angry parents, and I wind up the loser.

COUNSELOR: We need to develop a plan to help you.

_____
_____

CLIENT: What do you have in mind?

COUNSELOR: First, we need to make up a schedule of your week so we can identify the possible times you can devote to studying.

_____
_____

CLIENT: Well, I do have two free periods every day, but that's the time I spend in the cafeteria with my friends.

COUNSELOR: Let's try to put you on a self-reward system. What would you give yourself if you earned a reward?

_____
_____

CLIENT: Well, I like to socialize.

COUNSELOR: Would you, or can you, use this as a reward?

_____
_____

CLIENT: Yes, I think so.

COUNSELOR: Let's develop a study schedule for you for this coming week. Then we need to see how

_____
_____

we can use your social contacts
as a reward for studying when
you follow it.

CLIENT: Well, if I study the first
free period, I'll socialize the second one.

COUNSELOR: And, if you don't
study the first period?

_____

_____

CLIENT: I have to give it a try. I
have to improve or it's good-bye
for me.

CASE 8: Jane is a widow who has three children who live with her. She referred herself to your agency because of her lack of assertiveness; she appears
anxious and rather sad. She feels her lack of assertiveness is causing her to function poorly at work and at home.

Excerpts and Responses   *Ratings and Reasons*

COUNSELOR: What do you do
when someone cuts in front of
you in a line, say, at the
supermarket?

CLIENT: I get real mad, but I never
say anything.

COUNSELOR: So you do believe
that you should say something.

CLIENT: Yes, I know I should.

COUNSELOR: What prevents you?

CLIENT: I don't want to cause a
commotion.

COUNSELOR: So, on the one hand,
you want to assert yourself and
speak up for your rights, but on
the other hand, you have this
fear that prevents you from doing
so.

CLIENT: That's the dilemma I always have, and I get paralyzed.

COUNSELOR: It's not at all unusual to have these two feelings. But we have to change the balance and have your assertiveness become the more dominant one.

CLIENT: That's easy to say, but I always get paralyzed.

COUNSELOR: Learning to express your anger in any situation is hard. Perhaps it's related to your ability to express any of your feelings.

CLIENT: Well, I don't express the way I feel about things very often.

COUNSELOR: Okay, let's first work on having you learn to express your positive feelings.

CLIENT: What do you have in mind?

COUNSELOR: For the next week, I would like you to try to express your positive feelings to one of your children at least once a day.

CLIENT: I think I can do that.

Communication Exercises

The following exercises have been designed to enhance your problem-solving skills. Each exercise presents some critical information about a client that was obtained during an initial counseling session. These cases should be role-played in a triad. When you role-play as the counselor you should involve the client in selecting an appropriate goal for counseling, select the appropriate intervention strategy, and apply the strategy. Try to use different approaches for each case.

CLIENT 1: Robert is a 13-year-old junior high school student who is not doing very well in school. He is of average height and weight for his age, has a history of good health, and comes from an intact working-class family. He has one older brother and four older sisters; both parents work outside the home.

The assistant principal referred Robert and indicated that he had been accused by the children of stealing money in small amounts from them. There is no proof for this accusation; it is their word against his. Robert is reticent about volunteering information, but seems comfortable when allowed to relax in the counselor's office. He claims everyone picks on him—his family, his classmates, and his teachers—and no one really likes him or trusts him. His academic failures bother him, but he feels he cannot win. He does not appear to have a friend in the school.

CLIENT 2:   Mary, who is 21 years old, recently came to the United States. She lives with an older sister, her brother-in-law, and their four children. She had an isolated, disciplined childhood, and her parents were extremely protective of her because of a history of epilepsy. She has suffered from petit mal attacks since she was 5 years old. She has another older sister who is also married and lives in the same community. Both of her sisters are college graduates and both are employed in professional occupations; their husbands are both employed in midmanagerial civil service positions.

Mary reports that she gets along very well with her family. However, she feels that she is her sister's housekeeper and baby-sitter. She reports having a poor social life, and that she would really like to continue her education and become economically independent. She says that she had a high school average of 92 and loves to read and do crossword puzzles. She admits that she has never had any close friends except for her two sisters. She is active in her local church. The client talks rapidly and is highly verbal.

CLIENT 3:   Ed is a 53-year-old self-made man who has just lost his job in management with a prominent advertising agency where he worked for 31 years. He appears younger than his stated age, is well dressed, and has the bearing of a very accomplished, successful man. His wife has been moderately successful in the bank where she is employed. They have no children.

Ed came to talk to you because a friend, who is his physician, recommended he see a counselor because he was complaining about migraine headaches, backaches, and chronic fatigue even though he is apparently healthy. Ed admits feeling cheated after all the years of giving his best, but he sees no way that counseling can help him get his old job back or find a comparable position with another prestigious company. He only came to counseling because his physician friend urged him to talk out his feelings about his employment situation.

CLIENT 4:   Carol is a 16-year-old sophomore who is attending a public high school in a suburban community. Her parents are separated, and her father lives in a city 300 miles away. She lives at home with her mother and her sister, who is 14. Carol is not doing well in school; academically she is failing two subjects and barely passing the rest. Her deportment is considered awful: she is loud, boisterous, and insolent. She has a strong influence on other students.

As the counselor, you have recently encouraged Carol to stop by for a chat. During the initial interview (the chat) Carol put on a tough-girl act, but was open and sincere underneath. She began to relax. It soon became apparent that Carol and her sister are alone a great deal of the time. Her mother is out of town at work about three nights a week. According to Carol, her mother drinks a considerable amount of liquor, and Carol thinks that is the main reason her parents separated. Carol has no other immediate family in town; however, her father's sister lives in an adjoining community.

CLIENT 5:   Al is a 32-year-old college graduate who has just completed a master's degree in finance. He comes from a prominent family that is active in their community, which is several hundred miles away. Al has never worked in the business field. In college he was active in the college theater group and had several acting jobs in the off-Broadway theater and in summer stock. During a lull in his acting, he took a job in construction and had an accident, which caused a chronic lower back problem. As a result of some vocational counseling, Al went to graduate school and has just finished his course work. He has a good grade point average.

Because of his lower back problem, Al is reluctant to accept a job that will require sitting at a desk 8 hours a day. He is willing to relocate but does not want to return to his hometown to be his father's alter ego.

CLIENT 6:   Patricia is a 14-year-old student in high school. She lives at home with her parents and three siblings, sisters ages 16 and 28 and a brother age 10. Patricia is in good health, currently has a "B" average in her school work, and enjoys art and music.

The client came to the counseling center because she had difficulty relating to her father. According to her, he has a mental problem. He makes unrealistic statements, verbally degrades her when she disagrees with him, and embarrasses her with his bizarre behavior. The client is very sensitive, has a limited number of friends, and bases her self-worth on what others say and do. Patricia wants to "make something of herself" and feels she needs to learn how to cope more effectively with her father.

CLIENT 7:   Peter is 19 years old and a freshman at the university. He is the first member of his family to finish high school and attend college. He has had an excellent high school education and very good grades but now reports feeling strange and guilty. He lives at home, and his family is emotionally supportive of his going on to college. He senses a feeling of alienation and distancing, and a lack of understanding.

He is attracted to the humanities and is thinking of majoring in French or French literature. His family thinks he is not being very sensible and that he should think about a more practical major. Someone told him to stop by the university counseling center and pick up some literature on college majors and occupations. That is why he came to you. It was easy to establish a relationship with Peter. His presenting problem seems to be the real concern.

CLIENT 8: Martha is a 78-year-old widow who is a resident in an adult home. She worked until she was 70 years old as a secretary to the president of an industrial firm. She lost her right leg 6 months ago because of vascular problems. She has a married son who visits her once a week.

She was referred to you because she has complained constantly since arriving in the home 2 months ago. During the initial interviews Martha was quite talkative. She revealed that she was quick to have opinions of others, did not like the home, and was angry at her son because of this placement. Martha is a bright lady who is aware of the current events of the world, and she loves to read and discuss contemporary literature.

CLIENT 9: Kenneth is 19 years old and finished high school 6 months ago. He is the middle son in a family of five boys. Both parents work and are in the lower income range. His two older brothers do not live at home, and he and his younger brothers live in a group home. He has lived in this home for 6 years, and his two brothers were placed in the home 3 months ago. All three were placed in the group home by the court because they needed closer supervision.

During the initial interviews Kenneth was fidgety and distracted—he chewed his fingernails and tapped on the table. He has worked in the same unskilled job for the past 5 years. He manifests poor communication skills, a poor concept of time, and little sense of responsibility or independence. He was willing to come to see you because he thought he might be able to get a better paying job through your office.

CLIENT 10: Marilyn is a 27-year-old homemaker with two elementary-school-age children. She recently lost her job as a salesclerk in a discount department store and was self-referred to your agency because of her desire to seek another job. During your initial interview with Marilyn, you discovered that she was fired from her job because of a drinking problem and that she is having serious marital problems.

Marilyn is responsive and cooperative but is generally reactive rather than proactive. She is anxious to continue with the counseling.

CLIENT 11: Anne is a 17-year-old junior in high school. She has a "B+" average in her course work and is on the honor roll periodically. On standardized tests she always scores in the 99th percentile. She lives at home with both parents and a younger sister, age 15. She is tall for her age and in good health.

Anne came to see you because of her lack of assertiveness. She says that every time she goes to the supermarket someone manages to get ahead of her in the checkout line. She has few friends.

CLIENT 12:   Dan, a handsome 26-year-old, works as a computer specialist in a large corporation. He is married to Eileen, an attractive 23-year-old who works for the same company. Both of them seem happily married, are in good health, and plan to have children and a house in the suburbs within a few years. Dan's parents were divorced when he was 15, and his father remarried a year later. His two older sisters are now divorced and have returned home, each with a young daughter, to live with their mother, who always has a boyfriend but never remarried. Dan's father has young children and what Dan calls an "ideal marriage."

Dan has come to the counseling office because he is touched by nightmares that always focus on losing Eileen in some tragic way. He also wonders what is the matter with his family because no one is happily married. Everyone gets dumped. He realizes he is upsetting his own relationship with Eileen because of his restlessness and his fear of losing her.

CLIENT 13:   Doreen is a 16-year-old underachieving tenth-grade student who has extremely poor grades and a poor attendance record. She lives at home with her mother and two younger sisters and an older brother. She is of average height and weight and is in good health.

Doreen was self-referred. She came to talk to you because her mother wanted to send her to live with relatives out of town. During the initial interview Doreen revealed that she has little communication with members of her immediate family. She also revealed that she has few friends, is apathetic to school work, and has a poor self-image. Although she's quiet, Doreen appears to be comfortable in the counseling office.

CLIENT 14:   Charles is 50 years old and employed in an upper-level managerial position. He is married and has two adopted children, who are in junior high school. His wife is an elementary school teacher. He comes from a large family. His father died when he was very young, and he grew up in an economically poor home. Charles has a master's degree and is a CPA. His employment history is stable, and he has worked for the same firm for 20 years.

Charles is pleasant, friendly, and very willing to express himself. He came to the clinic because he is very unhappy about his marital situation. Although he wonders if his feelings of dissatisfaction are a symptom of a midlife crisis, he reports considerable emotional, physical, and social distance from his wife. Both of them are committed to the well-being of the children. However, he reports that "the home is not a sharing, warm refuge but a cold harbor out of the storm of life." Charles pursues his interests, his wife her's. There is cooperation within the family unit and no antagonism. But very rarely is anything done as a family unit.

CLIENT 15:   Susan is 25 years old, a magna cum laude graduate of college, teaching in a private elementary school, and pursuing a master's degree in education. She is of average height and weight, comes from a middle-class background, reports having had excellent health, and was married 6 months ago. Her husband is an accountant and is preparing for his CPA exams.

She came to the university counseling center because she is not pleased with either her professional work or her studies, and is convinced that she made a mistake in her career choice. The client was seen for two sessions. She is outgoing, friendly, and enjoys being in the center of things socially. However, she has also recently begun to question her relationships with others. She feels that although she has many acquaintances, she has no real friends, and that her relationship with her husband is marred by a lack of communication about the deeper issues of their relationship.

Exercises for Your Own Setting

List the types of problems that you encounter in your present setting, or in a setting that you expect to work in, and indicate the cognitively focused, affectively focused, or performance focused strategies that you would use with each of these problems.

SUMMARY

The problem-solving role communication skills outlined in this chapter briefly presented how you might employ a variety of cognitive, affective, and behavioral intervention strategies in the counseling process. A variety of client situations were provided to give you some initial practice in using these intervention skills. After reading about the various strategies outlined in Part 2 and practicing them in the exercises in the present chapter, you should have developed a basic understanding of them. You should be able to identify the types of intervention strategies used in a counseling situation and discuss the differences among the higher and lower level communication responses used by the counselor. Full mastery of the problem-solving skills presented in this chapter will require substantial supervised clinical practice. This supervised practice should provide you with the opportunity to study these strategies in more depth, to practice them with actual clients, and to receive constructive feedback from your professional peers and supervisors.

REFERENCES AND SUGGESTED READINGS

Baruth, L. G., & Huber, C. H. (1985). *Counseling and psychotherapy: Theoretical analysis and skills applications.* Columbus, OH: Merrill.

Burks, H. M., & Stefflre, B. (1979). *Theories of counseling* (3rd ed.). New York: McGraw-Hill.

Corey, G. (1991). *Theory and practice of counseling and psychotherapy* (4th ed.). Pacific Grove, CA: Brooks/Cole.

Cormier, W. H., & Cormier, L. S. (1991). *Interviewing strategies for helpers: Fundamental skills and cognitive behavioral interventions* (3rd ed.). Pacific Grove, CA: Brooks/Cole.

Corsini, R. J., & Wedding, D. (Eds.). (1989). *Current psychotherapies* (4th ed.). Itasca, IL: F. E. Peacock.

George, R. L., & Cristiani, T. S. (1990). *Counseling: Theory and practice* (2nd ed.). Englewood Cliffs, NJ: Prentice-Hall.

Gilliland, B. E., James R. K., & Bowman, J. T. (1989). *Theories and strategies in counseling and psychotherapy* (2nd ed.). Englewood Cliffs, NJ: Prentice-Hall.

Gladding, S. T. (1988). *Counseling: A comprehensive profession.* Columbus, OH: Merrill.

Hansen, J., Stevic, R. R., & Warner, R. W. (1986). *Counseling: Theory and process* (4th ed.). Boston: Allyn & Bacon.

Ivey, A. E., Ivey, M. B., & Simek-Downing, L. (1987). *Counseling and psychotherapy* (2nd ed.). Englewood Cliffs, NJ: Prentice-Hall.

Kanfer, F. H., & Goldstein, A. P. (1986). *Helping people change* (3rd ed.). Elmsford, NY: Pergamon Press.

Patterson, C. H. (1986). *Theories of counseling and psychotherapy* (4th ed.). New York: Harper & Row.

Shilling, L. E. (1984). *Perspectives on counseling theories.* Englewood Cliffs, NJ: Prentice-Hall.

Name Index

Subject Index

DATE DUE	
FEB 13 2001	
MAR 19 2001	